OIL AND GAS PRODUCTION

Corrosion Control

Edited by
Charles Kirkley

Published by
Petroleum Extension Service
Division of Continuing Education
The University of Texas at Austin
Austin, Texas

1982

Published 1982. Third Impression 1998
Printed in the United States of America

Catalog No. 3.30110
ISBN 0-88698-110-7

Table of Contents

Foreword . . . v
Acknowledgments . . . vii
How to Use This Manual . . . ix

1. The Corrosion Process . . . 1
   - Objectives . . . 1
   - A Definition of Corrosion . . . 2
   - Metals and Their Properties . . . 2
   - Electrolytes . . . 2
   - The Metallic Circuit . . . 3
   - Corrosion Cell Strength . . . 4
   - Corrosion Cell Size . . . 5
   - Types of Corrosion Cells . . . 6
   - Self-Test . . . 11
2. Common Corroding Agents . . . 13
   - Objectives . . . 13
   - Troublesome Corrodants . . . 14
   - Brine . . . 14
   - Carbon Dioxide and Other Acid-Forming Compounds . . . 14
   - Hydrogen Sulfide . . . 15
   - Oxygen . . . 15
   - Soil Moisture . . . 16
   - Bacteria . . . 16
   - Combinations of Corroding Agents . . . 17
   - Self-Test . . . 18
3. Detection and Measurement . . . 21
   - Objectives . . . 21
   - General Considerations . . . 22
   - Sampling and Chemical Analysis . . . 22
   - Testing with Metal Coupons . . . 22
   - Electrical Resistance Measurements . . . 23
   - Linear Polarization Instruments . . . 23
   - Hydrogen-Measuring Instruments . . . 25
   - Current and Potential Measurements . . . 26
   - Testing for Anaerobic Bacterial Corrosion . . . 27
   - Radiographic Examinations . . . 27
   - Ultrasonic Devices . . . 27
   - Electronic Pipeline Inspections . . . 28

Electromagnetic Examination of Casing . . . 29
Caliper Surveys . . . 29
Measuring Electric Current in Casing . . . 30
Inspection Records . . . 32
Measurement Combinations . . . 32
Self-Test . . . 33

4. Methods of Corrosion Control . . . 35
Objectives . . . 35
Control Measures . . . 36
System Design . . . 36
Corrosion-Resistant Construction Materials . . . 36
Coatings and Linings . . . 39
Insulation . . . 41
Inhibitors . . . 41
Oxygen Removal . . . 45
Self-Test . . . 47

5. Cathodic Protection . . . 49
Objectives . . . 49
Effectiveness . . . 50
The Cathodic Protection Process . . . 50
Sources of Power . . . 50
Power Requirements . . . 51
Impressed Current Anodes . . . 51
Galvanic Anodes . . . 54
Flow Characteristics of Cathodic Protection Currents . . . 57
Protective Coatings with Cathodic Protection . . . 58
Determining Cathodic Protection . . . 59
Cathodic Protection Interference . . . 61
The Future of Corrosion Control . . . 63
Self-Test . . . 64

Glossary . . . 67
Self-Test Answer Key . . . 75

Foreword

Corrosion is an economic enemy of huge proportions. Its enormous annual cost defies accurate accounting, but the efforts to prevent or mitigate its destructive effects are more easily accounted for in monetary terms. These preventive measures occupy the best skills of thousands of corrosion engineers and technicians, and the materials and energy used to combat corrosion result in annual costs measured in tens of billions of dollars.

Most production organizations employ small groups to devote full time to preventing corrosion. These corrosion engineers depend on operating personnel to install and operate corrosion control devices, make observations, participate in tests, and report results. So although an exhaustive knowledge of corrosion prevention may not be required of production workers, a better understanding of the subject will help promote the kind of cooperation that is essential to corrosion control efforts.

The purpose of this lesson is to provide a basic understanding of the corrosion process, some common corroding agents, methods of detecting and measuring corrosion, and various methods of corrosion control, with a special emphasis on cathodic protection. By applying this knowledge to operating procedures, production personnel can help reduce the immense cost of time and materials that corrosion damage inflicts on the petroleum industry.

Bruce R. Whalen
Publications Coordinator

Acknowledgments

The material contained in this lesson is largely representative of that taught in the PETEX School of Production Technology by Ray Wiederhold, Sun Production Company; Ray Bash, Cathodic Protection Service, Inc.; and Floyd Thorn, also of Cathodic Protection Service, Inc. Russell Brannon, a former Exxon Pipeline Company corrosion engineer who is also a founder and past president of the National Association of Corrosion Engineers, audited the presentations at the school and wrote the lesson.

Thanks go to Ray Wiederhold for reviewing the manuscript and suggesting content revisions, and also to Scott Smart, Petreco Division of Petrolite Instruments, for reviewing the manuscript and providing several helpful photographs.

Charles Kirkley
Editor

How to Use This Manual

The format of this manual includes a set of specific objectives for each section; at the end of the section is a competency self-test. To get maximum benefit from the manual, read the specific objectives carefully before studying the material in each section. As you study the material in the section, take notes, using the objectives as a guide to the most important parts.

When you feel that you have mastered the objectives, begin the self-test. Since it is a self-test, *you* must decide whether you should refer back to the material to answer the questions by determining how important that section is to your work. If you feel that you need to be very competent in an area, do not refer back until you have finished the test. This way, using the scoring points given at the beginning of the test, you can determine your percentage of competency. Score the test by using the corresponding key provided at the end of the manual.

1

The Corrosion Process

OBJECTIVES

Upon completion of this section, the student will be able to:

1. Define *reduction* and *oxidation.*
2. Explain the difference between *base metals* and *noble metals.*
3. Define *electrolyte.*
4. Name the components of a corrosion cell.
5. Describe the various chemical reactions associated with corrosion cells.
6. Define *anode* and *cathode.*
7. Explain why corrosion cell action is said to be *cathodically controlled.*
8. Name various types of corrosion cells and explain how they occur.

A DEFINITION OF CORROSION

When a metal comes into contact with an electrically conductive solution under certain conditions, an electrochemical reaction that changes the composition of the metal takes place. This reaction is called *corrosion.* A look at metallic properties and some basic chemical processes will help illustrate the fundamentals of the corrosion process.

METALS AND THEIR PROPERTIES

Most metals occur in nature in combination with other materials in the form of ores or salts. Metals are obtained from ores by *reduction,* a forcing of the metal's *ions* (charged atoms) to give up their charges and become metallic molecules. For this to take place, energy in the form of heat or electricity must be present. This process can be reversed. Metals that have been reduced tend to react with their surroundings and return to their earlier form when electric current is present. This reaction is called *oxidation.* Oxidation does not necessarily result in the formation of metal oxides; rather, it can be defined as the loss of *electrons* (e^-) (negatively charged atomic particles) by a constituent of a chemical reaction. The oxidation of iron (Fe) is represented in the formula

$$Fe^\circ - 2e^- \longrightarrow Fe^{++}.$$

Secondary reactions frequently occur in which the positively charged metal ions react with available negatively charged ions to form other compounds. In the formula below, iron ions react with hydroxyl (OH) ions to form ferrous hydroxide [$Fe(OH)_2$]:

$$Fe^{++} + 2(OH)^- \longrightarrow Fe(OH)_2.$$

These secondary reactions, though not a part of the oxidation process, may affect the rate at which oxidation occurs.

Metals differ in their tendencies to react or combine with other materials. The most reactive, *base metals,* are potassium, calcium, and sodium. The least reactive—gold, platinum, and mercury—are called *noble metals.* Iron, in the form of steel, is the metal of most concern to the producers of oil and gas. Compared with other metals, iron has moderate reactive tendencies. Depending on the circumstances, iron may remain unchanged for centuries, or it may react with its surroundings at such a rate that its useful properties are adversely affected after short periods of service.

ELECTROLYTES

The electrically conductive solution that must be present for corrosion to occur is called an *electrolyte.* When a metal comes in contact with an electrolyte and two or more points on the wetted surface of the metal have different electrical potentials, a continuous electrical path, or *corrosion cell,* is created between the areas with differing potentials.

Electrolytes conduct electricity because they contain ions. In oil and gas production, the most commonly encountered electrolytes are water solutions of salts, acids, or hydroxides. Pure water, by contrast, is a poor conductor of electricity. A few pure water molecules (H_2O) may be disassociated into positively charged hydrogen (H) ions and negatively charged hydroxyl ions, but the number of these ions is too small to allow pure water to conduct electricity. Absolutely pure water is a rarity; the purest available is referred to as *high-purity water.* Ionic solutions of salts, acids, and hydroxides are much more conductive and tend to accelerate corrosion cell action.

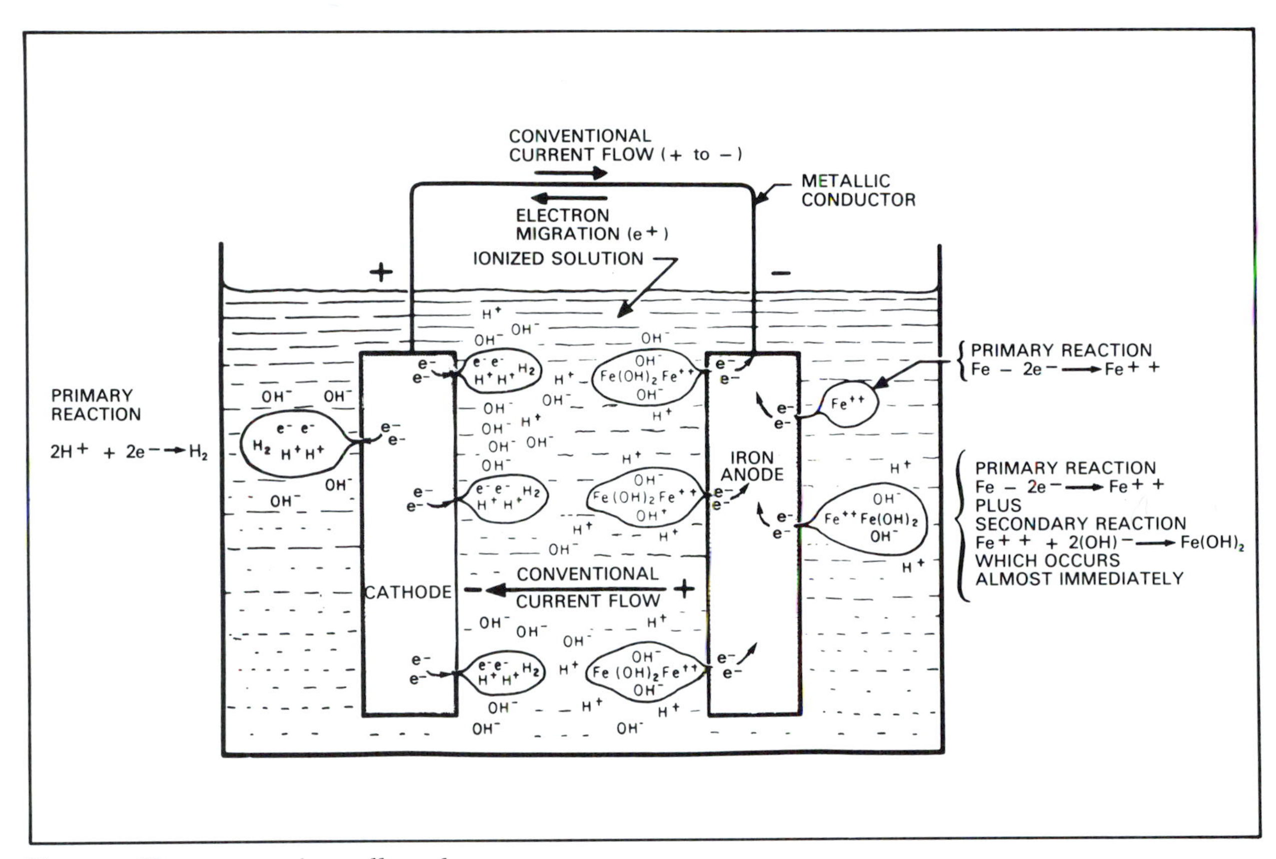

Figure 1. How a corrosion cell works

THE METALLIC CIRCUIT

Two metals linked in a corrosion cell illustrate typical reactions (fig. 1). The *cathode* in a corrosion cell is a point on the surface of the wetted metal which has a surplus of electrons and therefore carries a negative charge. When positively charged hydrogen ions come into contact with a cathode, they absorb the negative charges. The hydrogen ions are thus neutralized and converted into molecular hydrogen gas, as expressed in the formula

$$2H^{+} + 2e^{-} \longrightarrow H_2.$$

The loss of negative charges from the cathode lowers its electrical potential with respect to other parts of the metal. The cathode then picks up electrons that flow through the metallic path from the *anode,* or positively charged point in the cell. This electron flow creates a dearth of electrons at the anode; consequently, the metal gives up electrons and is converted into positively charged ions. This reaction, with iron as the metal, is symbolized in the formula

$$Fe - 2e^{-} \longrightarrow Fe^{++}.$$

At the anode there is almost immediately a secondary reaction in which the newly formed iron ions combine with readily available hydroxyl ions in the electrolyte, according to the formula

$$Fe^{++} + 2(OH)^{-} \longrightarrow FE(OH)_2.$$

The resulting ferrous hydroxide is insoluble in water and separates from the electrolyte. Without the iron ions, the cell continues to work without hindrance at the anode.

The essential parts of a corrosion cell may also be depicted as a triangle (fig. 2). Action in one part of a cell does not necessarily cause action in another area, but each element must be in place and begin functioning simultaneously. This situation changes, however, as the cell works.

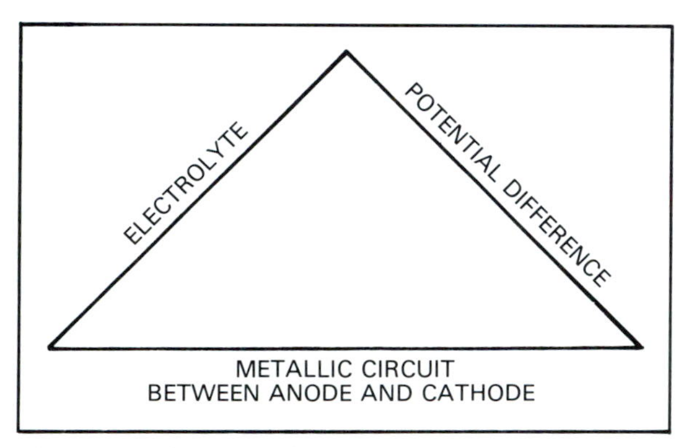

Figure 2. The corrosion triangle

At the cathode, one of two reactions may occur:

$$H^+ \longrightarrow \tfrac{1}{2}H_2 - e^-,$$

or

$$2H^+ + \tfrac{1}{2}O_2 \longrightarrow H_2O - 2e^-.$$

The first reaction occurs fairly rapidly in acid solutions, but very slowly in alkaline or neutral media. It can be speeded up by dissolved oxygen (O_2), as shown in the second reaction. Since the oxygen in solution is used up in the cathodic reaction, and the reaction rate depends on the rate at which oxygen reaches the metal surface by diffusion through the electrolyte, the rate of cell action is said to be *cathodically controlled.*

CORROSION CELL STRENGTH

The magnitude of strength, or driving force, of a corrosion cell is determined by the difference in electrical potential between the cathode and anode. This force is measured in volts (V). For a single metal, such as steel, which is exposed to electrolytes normally encountered in production operations, differences in potential rarely, if ever, exceed $\frac{3}{10}$V.

Corrosion cell strength can be influenced by the amount of material dissolved in an electrolyte. In most cells caused by different concentrations of electrolytes, the material exposed to the more concentrated solutions is usually anodic to other material. For this reason, well casings or tubing, when exposed to various geological formations, are more likely to be anodic in strata where brine, or salt water, is found.

Cathodes tend to be large in area compared to anodes. In the absence of oxygen, hydrogen deposited on cathodes as the result of cell action tends to remain as a polarizing film that resists additional hydrogen ions. The additional ions must seek points of entry at locations more remote from the anode until the voltage equals the current (I) times the resistance (R), according to Ohm's Law. Resistance, in this case, is created by the metallic path, the electrolytic path, and the polarization of the cathode and anode. This fundamental relationship for multiple cell action is described by Kirchoff's second law as

$$\sum emf = \sum IR.$$

Hence

$$E_c - E_a = IR_e + IR_m$$

where

R_e = the resistance of the electrolytic path;

R_m = the resistance of the metallic portion;

E_c = the effective (polarized) potential of the cathodic member of the couple; and

E_a = the effective (polarized) potential of the anodic member.

At this point, the cell action is stifled until something happens to reduce the polarization of the cathode or anode or both. This is usually accomplished when hydrogen reacts with oxygen at the cathode, according to the formula

$$2H^{+} + \frac{1}{2}O_2 \longrightarrow H_2O - 2e^{-}.$$

Once the $E = IR$ equilibrium is reached, further cell action depends principally upon the rate of hydrogen removal from the cathode. That rate, in turn, is largely controlled by the rate at which oxygen reaches the cathode. As noted earlier, the iron ions liberated at the anode quickly react with constituents of the electrolyte to form either insoluble compounds or soluble salts that migrate out into the electrolyte. Under certain conditions, corrosion products may form nonconductive, almost waterproof coverings that tend to stifle all cell action. These coverings usually form where multiple small cells occur, such as on metals exposed to mild atmospheric conditions. In many cases they do not restrict the further action of the cell.

Since cathodes tend to expand and crowd in on anodes, they may stifle weak cells in the vicinity of strong anodes. Anodic metal loss is concentrated in small areas, and deep corrosion pits form. This phenomenon causes the failure of structures in which the total metal loss is quite small.

Corrosion cells may act in parallel so that considerable metal areas are anodic and still larger areas are cathodic. Electrical currents produced by aggregations of cells acting in parallel may reach several amperes (amps) in size. These large, usually measurable currents are likely to occur on pipelines or well casings – structures that traverse areas with varying conditions of soil moisture, electrolyte strength, and oxygen concentration. Anodic and cathodic areas of such structures may be miles apart.

CORROSION CELL SIZE

Large cells such as those already described can cause large quantities of metal loss. Very large cells usually involve massive amounts of electrolyte like soil moisture or large bodies of water. Conditions often exist in which such cells can be set up, causing extensive damage to structures. These conditions are fairly easy to recognize, and steps can be taken to minimize and control them. As a matter of fact, such cells can be created intentionally as a method of corrosion control.

Corrosion cells can also be extremely small, as they are between adjacent grains of metal that have different surface potentials in an electrolyte. A cell can even be set up in a single droplet of electrolyte on a metal surface. The outer surface of the droplet would be exposed to oxygen in the air while the part in contact with the metal would have a lower concentration of dissolved oxygen, resulting in an oxygen-concentration cell.

Cell size is a matter of relative importance. Corrosion damage depends on the number and speed of small cell reactions and where they occur. Zinc, for instance, can be rapidly consumed when hydrogen gas evolves in acid, a reaction caused by the action of a myriad of small cells. As one anode is eaten away, another forms, until the metal is consumed or the acidity of the electrolyte is used up. A small cell that causes penetration of a thin-walled condenser tube may result in extensive damage and great expense.

It may seem strange that the small, often miniscule, driving voltages of corrosion cells can be so efficient in converting metals from the reduced to the oxidized state. However, the metallic paths offer little electrical resistance to the flow of the small currents involved. The voltage drop (or the *IR* drop, as expressed in the formula $E=IR$) caused by current flow is often negligible. The electrical resistivity of electrolytes is much higher than that of metals, but since the volume of available current paths is comparatively great, the total circuit resistance is low, and the cells are highly efficient.

TYPES OF CORROSION CELLS

Although a basic electrochemical process is common to all corrosion, corrosion problems can be classified into distinct types. Among the general causes of corrosion are oxygen concentration, galvanic action, stress, stray currents, erosion, and nonuniform surfaces of structural materials.

Oxygen-Concentration Cells

When metal surface conditions are uniform and the electrolyte contains varying amounts of dissolved oxygen, *oxygen-concentration cells* occur. Unlike other cells, the anodes (and the corrosion) form where oxygen concentration is lowest (fig. 3). Common examples of this phenomenon are nails driven into wood that has been soaked for long periods. The nail heads, exposed to oxygen in the air, are relatively unaffected, but the nail shanks just below are often narrowed to the point of failure. In oil and gas wells, surface structures such as flow lines and tanks, which have ready access to atmospheric oxygen, become strongly cathodic to downhole structures unless metallic paths between them are interrupted.

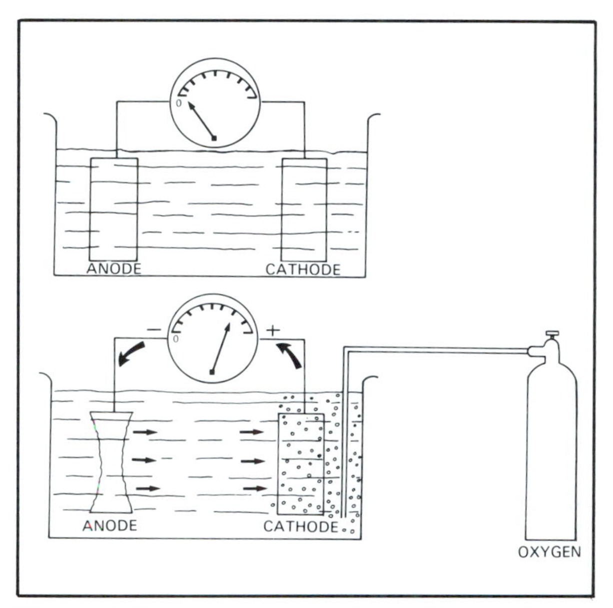

Figure 3. Oxygen-concentration cell corrosion

Electron flow in oxygen-concentration cells is in the same direction as in other cells. For this reason, oxygen-concentration cells may tend to interact with and intensify the action of other cells.

Galvanic Cells

It has long been recognized that certain metals, when used alone, have good resistance to corrosion. The same metals, when placed in contact with another metal in an electrolyte such as seawater, corrode rapidly. This reaction, brought about by the difference in electrical potentials of the dissimilar metals, was first utilized by the Italian scientist Luigi Galvani (1737–98) to construct battery cells that produced useful amounts of electricity. *Galvanic cells,* as these reactions came to be called, are the cause of many problems in oil and gas production structures.

The electrical potential of a galvanic cell is equal to the difference in potential values for the two metals in the cell. For any two metals connected in a galvanic cell, the anodic metal will be the one that is higher in

TABLE 1
Electromotive Series for
Various Metallic Elements

Metal	Potential (V)
Magnesium	-2.40
Aluminum	-1.70
Zinc	-0.76
Iron	-0.44
Cadmium	-0.40
Nickel	-0.23
Tin	-0.13
Lead	-0.12
Hydrogen	0.00
Copper	0.34-0.50
Silver	0.80
Platinum	0.86
Gold	1.36

the electromotive series; the cathode will be the metal lower in the series. Table 1 shows the electromotive series for various metals. Hydrogen appears in the series because, in electrochemical reactions, it acts like a metal. Its potential is expressed as 0 in this table, separating the more reactive base metals above it from the less reactive noble metals below. This series can be useful, but the information it provides is based on standards of metal purity, temperature, and electrolyte composition and strength. Such standards are rarely encountered in actual practice. A practical galvanic series, as shown in table 2, provides information on more commonly employed materials.

TABLE 2
Practical Galvanic Series

Metallic Element or Alloy	Potential (V)
Commercially pure magnesium	-1.75
Magnesium alloy (contains significant amounts of aluminum, zinc, and manganese)	-1.6
Zinc	-1.1
Aluminum alloy (contains zinc)	-1.05
Commercially pure aluminum	-0.8
Mild steel (untarnished)	-0.5 to -0.8
Cast iron	-0.5
Lead	-0.5
Brass, bronze, or copper	-0.2
High-silicon cast iron	-0.2
Mill scale on steel	-0.2
Carbon, coke, or graphite	0.3

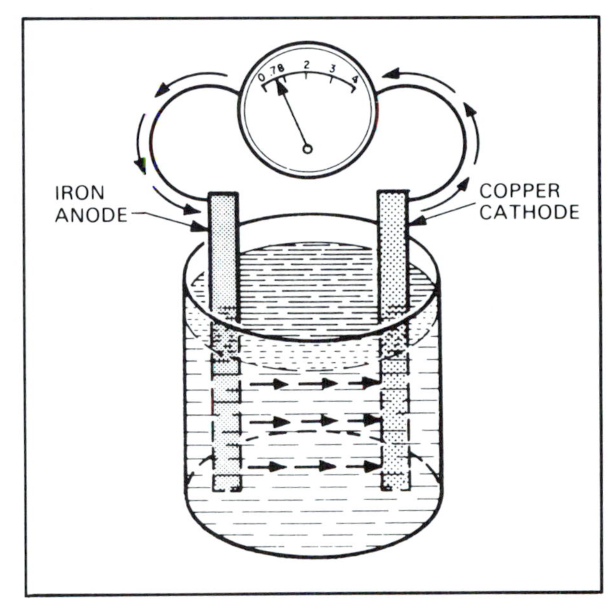

Figure 4. A copper-iron cell

The effects of galvanic action on steel are now so well known that harmful metal combinations, such as steel and copper, have practically been eliminated from production equipment that could be exposed to electrolytes. A copper-iron cell will produce ¾-1 V, depending on the purity of the elements, and the iron will corrode rapidly (fig. 4). In an iron-magnesium cell, the iron will be cathodic and will be protected by the magnesium. The voltage produced will be relatively high (fig. 5). Combinations beneficial to steel, such as steel and zinc, are widely used. However, oil and gas production often requires that grades of steel with

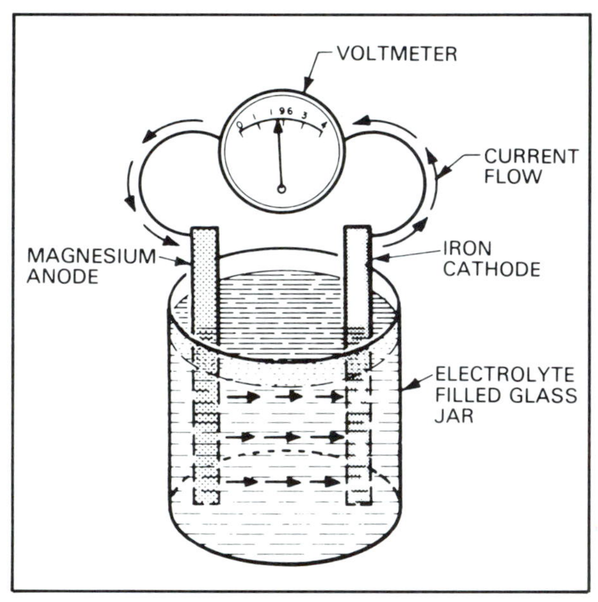

Figure 5. An iron-magnesium cell

certain properties – high strength, hardness, flexibility, or corrosion resistance – be used in conjunction with ordinary grades. With such combinations galvanic corrosion may occur. Alloying, hardening, quenching, or other modifications of steel may also affect galvanic potential. Through experimentation, the behavior of metals under various conditions of exposure can be determined. Specialists are usually needed to select the most compatible combinations of metals.

Stress

Stress can also be a factor in the formation of corrosion cells. The more highly stressed portions of the metal are anodic. Stresses may be locked in as a result of forming procedures, or they may result from service conditions. The production of oil and gas from greater depths requires metals with greater strength, hardness, erosion resistance, and heat tolerance. These desired properties may be obtained by alloying steel with carbon or metals such as nickel, chromium, molybdenum, and columbium or by heat treating, quenching, or tempering. Many of these altered metals have outstanding strength and corrosion resistance, but they may be subject to stress-corrosion cracking. This type of corrosion occurs along the grain boundaries of the metal and causes little metal loss. Because the process is not readily apparent, it may continue until the bonds between a sufficient number of metal grains are destroyed and the metal fails under stress. Since it is difficult to detect, stress corrosion may be particularly damaging.

Stray-Current Cells

Sometimes the electric current providing the driving force for a cell comes from an outside source. Metal may be immersed in an electrolyte that is carrying direct current from sources not electrically connected. Because of its greater electrical conductivity, the metal tends to collect current at points nearest the current source and discharge it back to the electrolyte at points more remote from the source (fig. 6).

When streetcars and interurban trains were in wide use, problems with buried structures were often encountered. Direct current was to be supplied to operating cars through a cable and returned to generating stations through the rails. The rails were often poor conductors, and the current "strayed" into the soil moisture and returned as electrolytic current. These cells came to be called *stray-current cells.* In oil field production, elongated or buried structures such as pipe and guy-wire anchors are naturally susceptible to stray currents, which still cause serious corrosion.

Ironically, the most common sources of stray currents are cathodic protection units. These currents, introduced into the ground or water to control corrosion, can become stray currents to foreign structures in their paths.

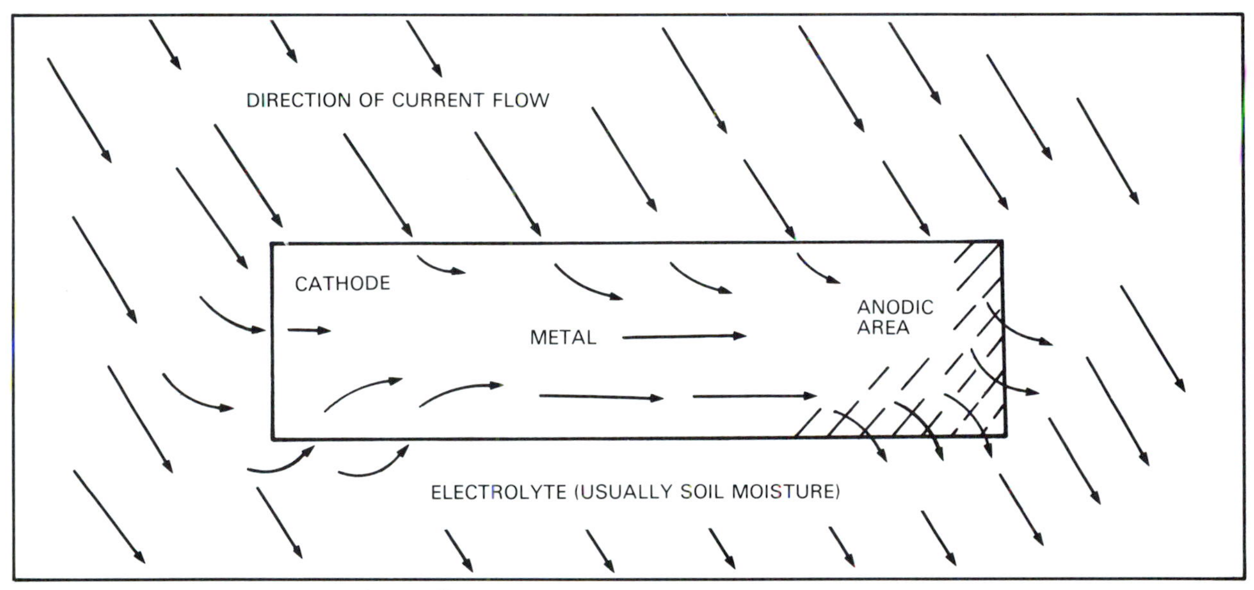

Figure 6. Stray current corrosion cell

Erosion

Some production fluids have particles of sand, corrosion scale, or other materials that tend to abrade or scour metal surfaces when liquid flow is turbulent. Even when the scouring action is not enough to remove metal mechanically, it may remove surface deposits such as hydrogen on the cathodes of galvanic cells, stimulating the action of the cells. Turbulent flow of fluids without abrasive particles can also have this effect. Protective coatings may be worn away, allowing for corrosion of the exposed metal below. This process is similar to that of windblown or tire-thrown sand removing paint from automobiles, which causes the consequent rusting of metals. This action occurs most frequently in piping, especially at bends, elbows, orifices, and other fittings where the direction of flow is changed or where flow turbulence is increased. The rubbing action of sucker and piston rods against tubing can also cause excessive wear and reduce the effectiveness of some corrosion control measures like protective coatings.

Nonuniform Surfaces of Structural Materials

Materials other than metals can also form corrosion cells with steel. Carbon, in the form of graphite, coke, or iron carbide, is cathodic to steel. Corrosion cells may be set up when grains of iron carbide in steel are exposed at the steel surface (fig. 7). New steel in ingot form is heated to red or white heat, then rolled, forged, or otherwise formed into useful shapes. As the steel cools, oxides form on the steel surface as a thin, dense, tightly adhering scale known as *mill scale.* Mill scale is electrically conductive and lower in the galvanic series than steel, so it becomes cathodic to its own steel base. Mill scale is usually cracked or partially removed during handling and construction, so portions of the steel are exposed. The action of such corrosion cells can cause early failure of production equipment, especially casing pipe and line pipe.

Differences in the condition of steel surfaces can also set up corrosion cells. A common example of this is when a section of

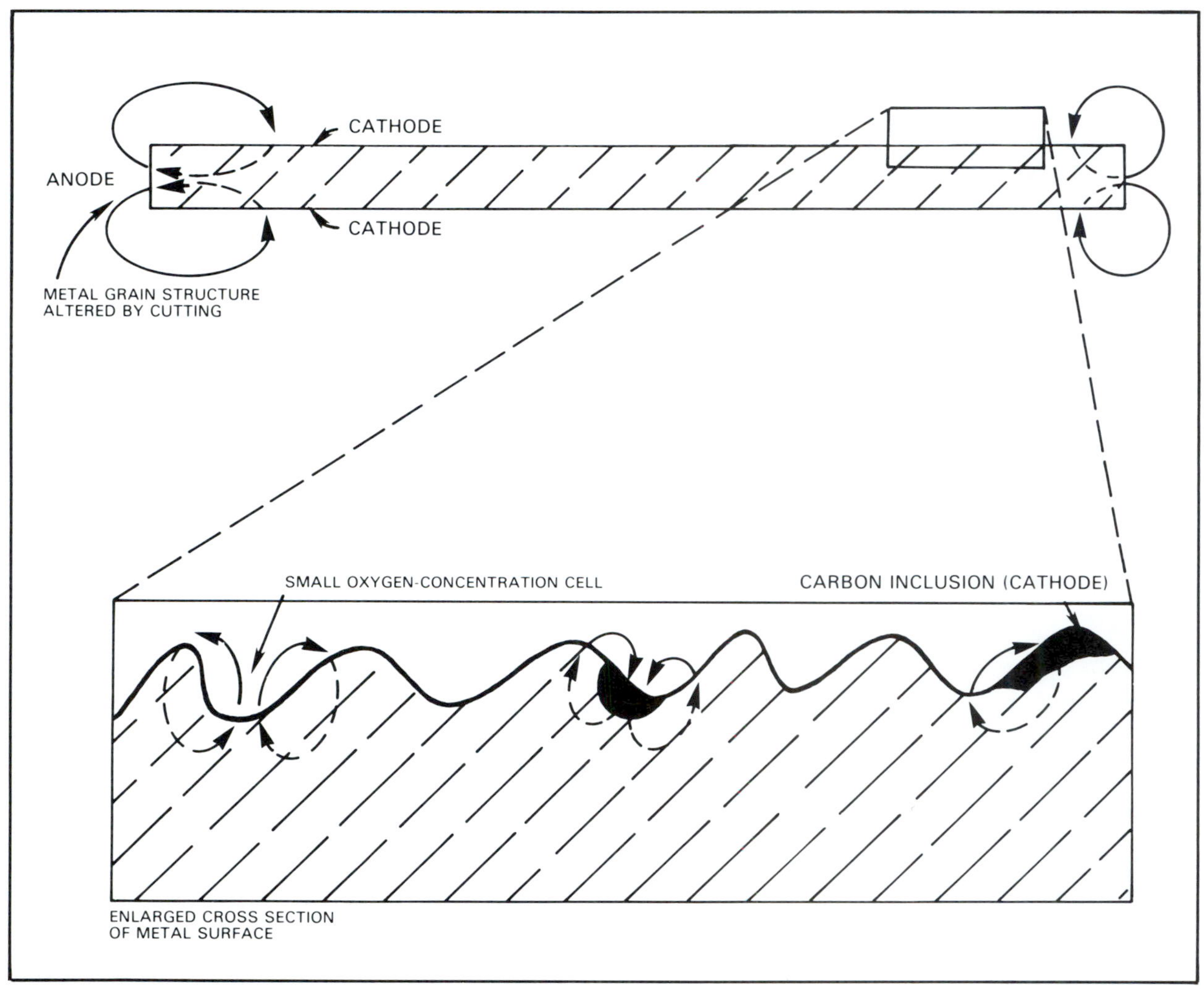

Figure 7. Carbon inclusion cell

new pipe is inserted into an existing pipeline, casing string, or tubing string. The new pipe is anodic to the old pipe and is often subject to fast corrosion attack. Removing corrosion products from portions of a steel surface may make the cleaned surfaces anodic to the rusted surfaces.

Differences in grain structure of metals may also cause corrosion cells to develop. Pipe that is produced by forming steel plates into tubes and welding the longitudinal seams may corrode in the heat-affected zones of the weld, unless the grain structure in those areas is normalized (fig. 8).

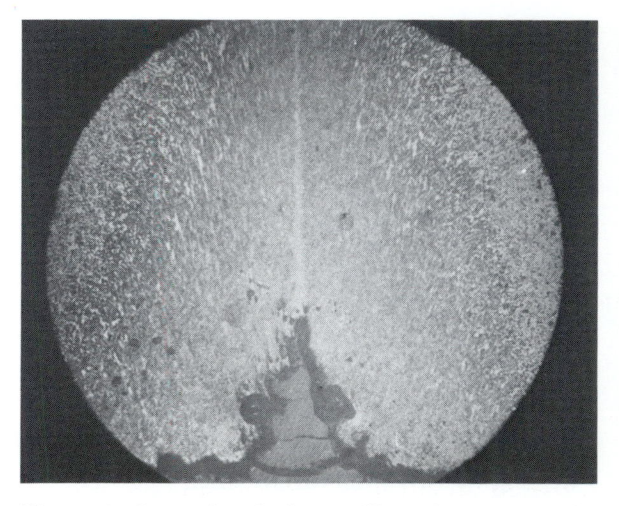

Figure 8. Corrosion in heat-affected zone at weld

Self-Test

1: The Corrosion Process

(Multiply each correct answer by five to arrive at your percentage of competency.)

Fill in the Blank

1. Metals are obtained from ores by ______________________, a process of giving up ions and becoming metallic molecules.

2. The process by which metals react with their surroundings and return to their natural state is called ______________________.

3. Because of their highly reactive tendencies, potassium, calcium , and sodium are referred to as _____________ metals.

4. Because of their poor reactive tendencies, gold, platinum, and mercury are referred to as _______________ metals.

5. When a metal comes in contact with an electrolyte and two or more points on the wetted surface of the metal have different electrical potentials, a ____________________ _______ is created between the areas with differing potentials.

6. The most commonly encountered electrolytes are water solutions of _____________, _____________, and ______________________.

7. In a corrosion cell, electrons flow from the positively charged _____________ to the negatively charged _________________.

8. Since oxygen in solution is used up in the cathodic reaction, and the reaction rate depends on the rate at which oxygen reaches the metal surface, the rate of corrosion cell action is said to be ___________________________ ______________________.

True or False

9. ______ When the $E=IR$ equilibrium is reached, corrosion cell action is stifled until cathodic or anodic polarization is reduced.

10. ______ Production structures that traverse areas with varying electrolytic conditions are not usually subject to large electrical currents produced by aggregations of corrosion cells.

11. ______ The corrosion of zinc in acidic electrolytes proves that large corrosion cells do the greatest damage.

12. ______ Cathodes tend to be smaller in area than anodes.

Matching

13. ______Oxygen-concentration cells
14. ______Stress corrosion
15. ______Erosion corrosion
16. ______Galvanic cells
17. ______Stray-current cells

A. Caused by electrical currents between dissimilar metals.
B. First noticed in the use of streetcars.
C. Caused by fluid flow or rubbing action of production equipment.
D. Caused by forming procedures or service conditions.
E. May be determined by exposure to air.

2

Common Corroding Agents

OBJECTIVES

Upon completion of this section, the student will be able to:

1. Name several of the corroding agents that commonly cause problems in oil and gas production.

2. Give several reasons why brine is such a troublesome corrodant.

3. Define *splash zone.*

4. Describe a few of the ways in which corroding agents contribute to the formation of oxygen-concentration cells.

5. Explain how the presence of carbon dioxide and hydrogen sulfide can create compounds that are harmful to production structures.

6. Explain the difference between *aerobic* and *anaerobic* bacteria and describe how bacteria affect corrosion.

7. Define *synergistic effect.*

TROUBLESOME CORRODANTS

The corrosion of most concern in oil and gas production is that which occurs inside piping, fittings, separators, treaters, tanks, and downhole structures. The outside surfaces of above-ground structures may also be subject to severe corrosion, but they are more readily available for inspection and maintenance. The most troublesome corroding agents are those which are produced along with the oil and gas. Among the most common corrodants are brine, carbon dioxide and other organic acids, hydrogen sulfide, oxygen, soil moisture, and bacteria.

BRINE

Brine is found along with crude oil and gas in producing formations. Some brine is almost always produced along with oil, and often with gas. The proportion of brine to oil may at first be negligible, but it increases with the field's productive life. Some wells end up yielding much more brine than oil.

Brine is often referred to as *water* or *salt water.* Oil field brines vary in composition, but in general they are water solutions of sodium chloride with additional ions. Positively charged *cations* such as calcium, magnesium, and hydrogen may be found together with negatively charged *anions* such as sulfate, bicarbonate, sulfide (from hydrogen sulfide), and hydroxyl. Concentrations of dissolved salts vary, but generally they are about 10,000 parts per million (ppm), or three times the salinity of seawater. These solutions, with their high proportions of ionizable salts and elevated temperatures, provide low-resistivity electrolytic paths for corrosion cells. Brine also provides some of the elements for other corroding agents.

Seawater may be classified as brine. Since it is usually plentifully supplied with oxygen, it is an aggressive corroding agent. With more and more production offshore, seawater has become a corrosion threat to structures that are constantly submerged or intermittently wetted by tides, waves, or spray. Drilling and production platforms, pipelines carrying oil and gas ashore, underwater storage tanks, boats, barges, and ships are among those structures exposed.

The most serious corrosion occurs on equipment that is regularly wetted, but not completely submerged. Control measures in these *splash zones* are few and expensive. The metal areas exposed are usually exterior surfaces, so corrosion on submerged portions can be controlled by cathodic protection. Steel pipes that are continuously and completely filled with saturated brine are not subject to serious oxygen corrosion, because, as the concentration approaches saturation, the solubility of oxygen decreases.

CARBON DIOXIDE AND OTHER ACID-FORMING COMPOUNDS

Carbon dioxide (CO_2) and other acid-forming organic compounds may be present in crude oil and gas. Sometimes wells producing sweet crude oil or gas condensate where corrosion would not be expected suffer corrosion from these acids. Such reactions are typified by the equations

$$CO_2 + H_2O \rightarrow H_2CO_3$$

and

$$H_2CO_3 + Fe \rightarrow FeCO_3 + H_2.$$

Since carbonic acid (H_2CO_3) is only slightly ionized and therefore a weak acid, it would not be expected to cause much corrosion. Its corrosivity has been found to be proportional to its partial pressure, as expressed in

TABLE 3
Corrosivity of Carbonic Acid
Relative to Partial Pressure

Partial Pressure (psi)	Degree of Corrosion
0–10	Mild
10–30	Moderate
Over 30	Severe

table 3. For example, if it is assumed that wellhead flowing pressure is 1,000 pounds per square inch (psi) and carbon dioxide concentration (determined by chemical analysis) is 5 percent, then the partial pressure of carbon dioxide would be calculated as follows:

$$1{,}000 \times 0.05 = 50 \text{ psi.}$$

Since the pressure is about 30 psi, severe corrosion can be expected.

In some cases, especially in gas-condensate wells producing from deep formations where pressures and temperatures are high, a little corrosion at critical locations can result in tremendous losses. In deep wells, production tubing strings are so long and heavy that their upper portions are stressed nearly to allowable strength limits. Temperature and pressure near the wellhead are lower than at formation level, causing some condensation of water containing dissolved carbon dioxide. Corrosion cell acitivity is enhanced by the high temperature and pressure, the acidity of the carbonic acid in the nearly pure condensed water, the stressed condition of the metal, and the turbulent flow of gases and liquids that sweeps away the polarizing films of hydrogen on the cathodes of cells. Anyone familiar with production can visualize the consequences of dropping a long string of production tubing into a high-pressure gas-condensate well.

HYDROGEN SULFIDE

Many wells produce oil and gas containing hydrogen sulfide (H_2S). This compound reacts readily with iron in the presence of water (either brine cr condensed water) according to the formula

$$H_2S + Fe + H_2O \longrightarrow FeS + H_2O + H_2.$$

The important product of this reaction, iron sulfide (FeS), is a black, porous substance that is cathodic to iron and thus sets up galvanic cells that cause the corrosion to continue. Oxygen is usually absent at sites of such corrosion, so there is no tendency for corrosion products to oxidize into protective films.

OXYGEN

Oxygen accounts for a large proportion of all metal corrosion because of its strong inherent tendency to form metal oxides and because of its omnipresence in the atmosphere.

The presence of oxygen tends to increase the rates at which other corroding agents react. In a galvanic cell, oxygen combines readily with the polarizing hydrogen films on the cathode, depolarizing the films and hastening the corrosion process. Oxygen-concentration cells may result from oxygen-deficient areas becoming anodic to oxygen-saturated cathodic areas. An example of this would be pipe in varying soil conditions or areas beneath pipe that contains sand, scale, tubercles, or trash (fig. 9).

Oxygen can also react with constituents of water that is used to flood producing formations or with substances encountered within formations. Such reactions may produce solid powders or gelatinous masses that can plug formations and reduce the effectiveness of the flooding operation.

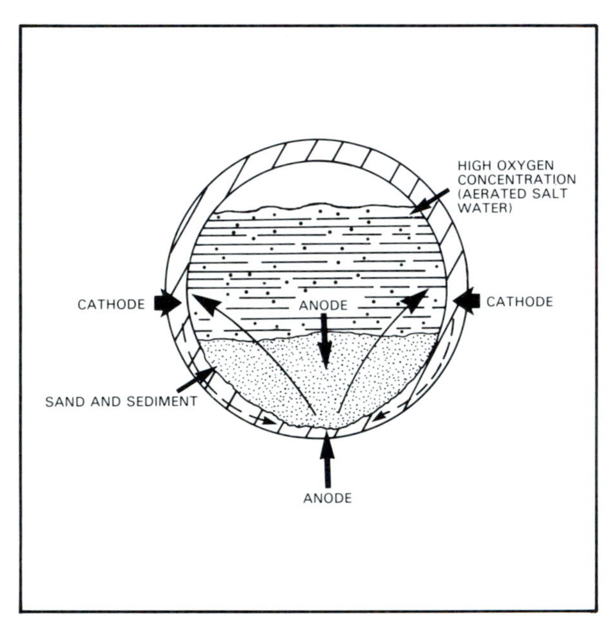

Figure 9. Oxygen-concentration cell caused by sediment buildup in pipe

But oxygen has a beneficial side, too. Oxidation products sometimes form protective coverings that resist further corrosion, like the pleasing blue green patinas on copper roofs, the dense oxide films on steel, and the often-invisible oxide films on aluminum. In fact, aluminum, a very reactive metal, would be a poor construction material were it not for its protective film, which is readily attacked by strong acids and alkalis. It is most stable in the 5–7 pH range. Even light rust on wrought iron has good resistance to atmospheric corrosion in some areas.

SOIL MOISTURE

Other than well casings, only a small proportion of production equipment is in contact with soil moisture. Flow lines and tanks are the surface structures principally exposed; production tubing is sometimes exposed downhole. The elongated nature of these structures brings them into contact with soils of differing characteristics, allowing for corrosion cell formation.

Corrosion failures in flow lines can be readily detected and located, and repair is fairly easy. Tank-bottom failures are more troublesome and expensive. Casing and tubing failures can be extremely disruptive and expensive.

BACTERIA

The life processes of tiny bacteria can contribute to corrosion. Bacteria can be divided into two classes, based on the way oxygen is utilized. This distinction is fundamental to the type and degree of corrosion damage caused by bacteria.

Aerobic Bacteria

Bacteria that flourish in the presence of oxygen, or *aerobic bacteria,* can produce slime or scum that accumulates on metal surfaces and creates oxygen-concentration cells. The sticky slime can enmesh sand particles (especially in areas of low flow velocity) and create deposits that are not easily removed by increased flow rates or other cleansing procedures. This type of bacterial corrosion may affect producing wells, water (brine) disposal systems, water flood projects, and various vessels, pipelines, and casings. Bacterial corrosion has been reported in such unlikely places as gasoline storage tanks and airplane fuel tanks.

Anaerobic Bacteria

Hydrogen sulfide corrosion is often caused by interaction of sulfate-reducing bacteria with sulfates in solution. These *anaerobic bacteria* live in the absence of free oxygen. They obtain the oxygen they need for their life processes by reducing the sulfate ion to sulfur, which combines with hydrogen ions in water to form hydrogen sulfide. Anaerobic bacteria may thrive and cause hydrogen sulfide corrosion in closed water systems where oxygen is effectively excluded and a good supply of sulfates is present in the water. The bacteria can also live in aerated systems under deposits of sand, rust scale, or other trash that excludes oxygen. These conditions also favor the creation of oxygen-concentration cells. Conditions favorable to the bacteria also occur in some soils, and pipes in those soils corrode.

COMBINATIONS OF CORRODING AGENTS

Quite often a combination of corroding agents is encountered and corrosion rates increase. Sometimes rates increase geometrically, as when one agent prevents the limiting action of another agent over a period of time or when the protective qualities of corrosion products are altered. Corrosion rates may also increase due to a *synergistic effect,* an incremental effect of two or more agents working together.

Self-Test

2: Common Corroding Agents

(Multiply each correct answer by six to arrive at your percentage of competency.)

Fill in the Blank

1. ____________ is often referred to as *water* or *salt water.*
2. The most serious brine corrosion occurs on structures that are located in ____________ ____________.
3. Wells producing sweet crude or gas condensate, where corrosion would not be expected, sometimes suffer corrosion from ____________ ____________.
4. Hydrogen sulfide reacts with iron in water to form ____________ ____________, which causes galvanic corrosion.
5. Because of its omnipresence in the atmosphere, ____________ accounts for a large proportion of all metal corrosion.
6. Oxygen-deficient areas become anodic to oxygen-saturated cathodic areas, forming ____________ cells.
7. Well casings suffer corrosion from ____________ ____________ more often than do other production structures.
8. ____________ bacteria live in the absence of free oxygen. ____________ bacteria flourish in the presence of oxygen.
9. Bacteria can cause ____________ ____________ corrosion in closed water systems where oxygen is excluded and a good supply of sulfates is present.
10. Corrosion rates may increase due to a ____________ effect, an incremental effect of two or more agents working together.

True or False

11. ______ The most troublesome corroding agents are those that are produced along with oil or gas.

12. ______ In the early life of a producing formation, the proportion of brine to oil may be great, but it decreases throughout the field's productive life.

13. ______ Oxidation products sometimes have beneficial uses.

14. ______ The corrosivity of carbonic acid is proportional to its partial pressure.

15. ______ The slime produced by anaerobic bacteria can accumulate on metal surfaces and create oxygen-concentration cells.

3

Detection and Measurement

OBJECTIVES

Upon completion of this section, the student will be able to:

1. Explain the importance of the corrosion rate in determining corrosion control measures.
2. Define *iron count.*
3. Describe the effect of organic acids in production fluids.
4. Describe how metal coupons are used to detect corrosion.
5. Name at least five different corrosion-measuring instruments and explain how they work.
6. Cite some of the basic problems in corrosion measurement and detection.
7. Define *fouling* and *passivation.*
8. Give the advantages and disadvantages of caliper surveys.
9. Explain the importance of keeping good records in measuring corrosion.

GENERAL CONSIDERATIONS

At every opportunity, the condition of production equipment should be visually inspected for corrosion damage. But most corrosion occurs on underground or undersea structures where periodic visual inspection is impractical, if not impossible. Detecting and monitoring corrosion inside pressurized vessels such as piping present major problems.

Once corrosion has been detected and located, the corrosion rate must be determined in order to decide which control measures, if any, to use and whether the expense of control or prevention is justified. Many times there are important considerations besides cost; safety, ecological concerns, and public relations are all important factors to be considered. The effects of previous corrosion on the present corrosion rate must also be determined. These measurements alone require a technology as complex and important as that of corrosion control. Among the commonly used methods of corrosion detection are sampling and chemical analysis, metal coupon testing, electrical resistance measurements, linear polarization measurements, hydrogen measurements, current and potential measurements, bacteria tests, radiographic examinations, ultrasonic inspections, electronic examinations of tubing and casing, caliper surveys, and records based on such measurements.

SAMPLING AND CHEMICAL ANALYSIS

Samples of oil, gas, brine, corrosion products, corroded articles, and test specimens must often be collected by field personnel and sent to laboratories for testing and analysis. The corrosivity of oil or gas should be tested at the earliest possible moment, when the first production from a newly discovered field is available, because by the time production samples have come in, steel in the exploratory well will have been exposed to possible corroding agents. Laboratory personnel and corrosion engineers need full information regarding field conditions affecting the samples, so all prescribed sampling procedures—proper handling, packing, shipping, and reporting of information—should be carefully followed.

Chemical analysis of produced oil or gas can sometimes determine the presence and the amounts of corroding agents or corrosion products. Analyzing these products can indicate the nature and the extent of the corrosion. The presence of iron compounds in solution or suspension is always a matter of concern. The *iron count,* a measure of iron compounds in the product stream, reflects the occurrence and the extent of corrosion and the effectiveness of any control measures that have been introduced.

Tests for iron may indicate corrosivity, but only bottomhole samples determine this for sure. The presence of dissolved gases such as oxygen, carbon dioxide, and hydrogen sulfide suggest possible trouble. Oxygen, though rarely found in producing formations, must be kept out of oil, gas, and the accompanying water so that corrosion rates will not be increased.

TESTING WITH METAL COUPONS

A widely used procedure for measuring corrosion rates is to expose coupons of metals to liquids or gases that might cause corrosion. The coupons are usually exposed to oil or gas streams flowing through piping under actual or simulated operating conditions. A coupon may be of the same metal as the piping (or whatever facility is used to determine the corrosion rate), or different metals may be exposed to determine which ones have satisfactory resistance to the corrodant involved.

Exposure periods for metal coupons may range from a few days to several years. Coupons of Hastelloy Alloy C, for example, have retained their mirror finish after thirty years of exposure to a marine atmosphere in a test that is still in progress. Ordinarily, coupons are examined at intervals. Total corrosion can be measured by weight loss; corrosion type (pitting or general attack) can be seen by visual examination; and the rate of penetration can be determined by examining pit depths or the thickness of the remaining metal.

It is difficult to place coupons so that they are subject to the same action that operating facilities are. Pipe nipples that become a part of the pipe being tested are often used. The coupon method has limitations, but it has produced a great deal of useful information.

ELECTRICAL RESISTANCE MEASUREMENTS

The corrosiveness of a medium such as oil or gas can be monitored by inserting a probe into a flowing stream or into a container of the medium and measuring the changes in the electrical properties of the probe as it corrodes, if it does. In some instruments, a suspected corrodant may be detected by inserting thin strips or wires of a susceptible metal into pipe or vessels. As the metal's cross-sectional areas are reduced, increases in electrical resistance can be measured, observed, and recorded. Such probes can be kept in place for extended periods as a check on inhibitors or other corrosion control measures. The readout instruments for such probes can be calibrated in terms of metal loss, and readings can be recorded on charts or telemetered to control locations if circumstances justify it. Other instruments operating on similar principles are available.

An important aspect of using measuring instruments is geometric placement of the probes so that they are exposed to representative proportions of the corrodant. For instance, if most of the corrosion occurs where brine settles to the bottom of a horizontal pipe, a probe located in the center or at the top of the pipe gives an incorrect indication.

Another concern is that probes may become *fouled,* or covered by deposits that affect corrosion rates. Probes are usually thin with small cross-sectional areas for increased sensitivity, and they need to be changed or examined at fairly short intervals. Probes are usually mounted in a way that permits replacement without interrupting the operation of the facility. Readout accuracy may sometimes be checked by the weight loss of the probes.

LINEAR POLARIZATION INSTRUMENTS

Instantaneous corrosion rates can be measured by means of an electrochemical phenomenon called linear polarization. Instruments using such techniques may be used in conjunction with computerized systems to monitor, analyze, and control corrective treatments. Polarization instruments change the electrical potential of a specimen corroding in a conductive fluid and measure the current required to make the change (fig. 10). So long as the change in potential does not exceed 10–20 millivolts (mV), the rate and nature of the corrosion reactions are not disturbed.

The amount of applied current required to cause the potential change is proportional to

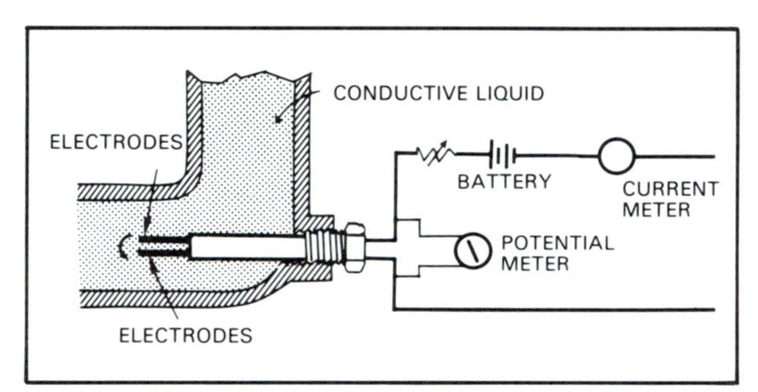

Figure 10. Two-electrode polarization instrument and diagram of installation and circuitry (Courtesy Rohrback Instruments)

the corrosion current. The following formula describes the relationship mathematically:

$$I_{corr} = K \frac{I_{app}}{\Delta E}$$

where

I_{corr} = corrosion current of specimen;

ΔE = change in electrical potential of specimen;

I_{app} = applied current required to change the electrical potential of the specimen by amount E; and

K = collection of constants.

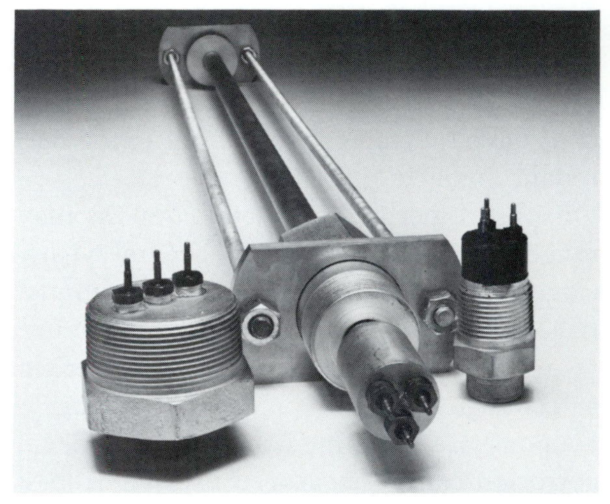

Figure 11. Probe assembly for linear polarization instrument

In practice, the constants can be compensated for by the instrument. The areas of both the specimen and the electrode, as well as the potential shift, can be set at fixed values so that the instrument can be calibrated to read the corrosion rate directly in mils per year (mpy).

The probe assembly (fig. 11), threaded for mounting into piping or vessels, is connected by a cable to the meter (fig. 12). The

Figure 12. Linear polarization instrument module and corrosion rate recorder.

Figure 13. Electrode assembly for linear polarization instrument

rod-shaped electrodes are of the metal to be tested (fig. 13). Corrosion rates are determined by measuring the applied current required to polarize the electrodes to a 20-mV difference in potential. The polarity of the applied current is then reversed, and the process is repeated.

Linear polarization techniques require that corrosion rates be measured in electrically conductive solutions, since current must flow from one electrode to another through the solution. Potential users of these instruments should seek advice from the manufacturer regarding an instrument's adaptability to a specific system. Such instruments cannot be used in oil or gas; even in intermittent oil, water, or gas flow, reliable results are difficult to obtain. Corrosivity of oil dispersed in water can be measured, but that of water in oil cannot. After installation, sufficient time must be allowed for the electrodes to reach equilibrium with their environments. The probes should be installed so that they can be readily removed for cleaning, inspection, and replacement. They can be short-circuited by conductive substances such as iron sulfide or coated by nonconductive solids such as paraffin, causing them to give erroneous results.

HYDROGEN-MEASURING INSTRUMENTS

A *hydrogen patch probe* can be strapped on to the exterior of a vessel, eliminating the need for opening the vessel to insert monitoring instruments. This probe consists of a powered, hydrogen-sensing cell and an instrument that instantly measures and records hydrogen reaction current. The probe provides a thirty-one–day record and can operate unattended for thirty days on two 12-V batteries. The hydrogen patch probe has some obvious advantages over other measuring tools: it is easily installed, and it measures the hydrogen produced by the corrosion of the vessels themselves, not foreign materials. However, it is a relative indicator of conditions, since not all of the produced hydrogen enters the steel.

Simpler instruments consist of closed steel tubes with pressure gauges and fittings for exposing the tubes to corrosive media. The particular type of steel for the tubes readily absorbs and transmits atomic hydrogen. The buildup of pressure in the tubes indicates hydrogen transmission. The tubes require periodic removal for cleaning with solvents and sandpaper to prevent fouling or *passivation,* the condition of being chemically inactive.

Steel containing atomic hydrogen in solution is brittle, and it may rupture from relatively small stresses. This condition, called *hydrogen embrittlement,* is most often associated with hydrogen sulfide corrosion.

Metals that suffer hydrogen embrittlement are not always damaged by it. If a metal does not crack or break while it is embrittled and it can be separated from the corroding agent, the dissolved hydrogen will dissipate and the metal will return to its usual condition of ductility and strength.

CURRENT AND POTENTIAL MEASUREMENTS

Very high resistance voltmeters (with mV ranges) or potentiometer-type voltmeters (which draw little or no current from the measured circuit) can be used with nonpolarizing electrodes to measure metal-to-electrolyte potentials. These measurements determine whether the metal is cathodic or anodic to the electrolyte. Nonpolarizing electrodes are used to contact the electrolyte and complete the measuring circuit. The electrodes consist of high-purity metal in contact with a saturated solution of one of the metal's own salts. The solution and the electrolyte under test make contact through a porous portion of the solution container. This portion is porous enough to become saturated with the solution without allowing significant leakage.

The most commonly used nonpolarizing electrode is the copper-copper sulfate electrode, often called more simply the copper sulfate electrode (fig. 14). These electrodes are also called *half-cells.* The other half of a complete electrolytic circuit for such measurements is the area where the structure contacts the electrolyte.

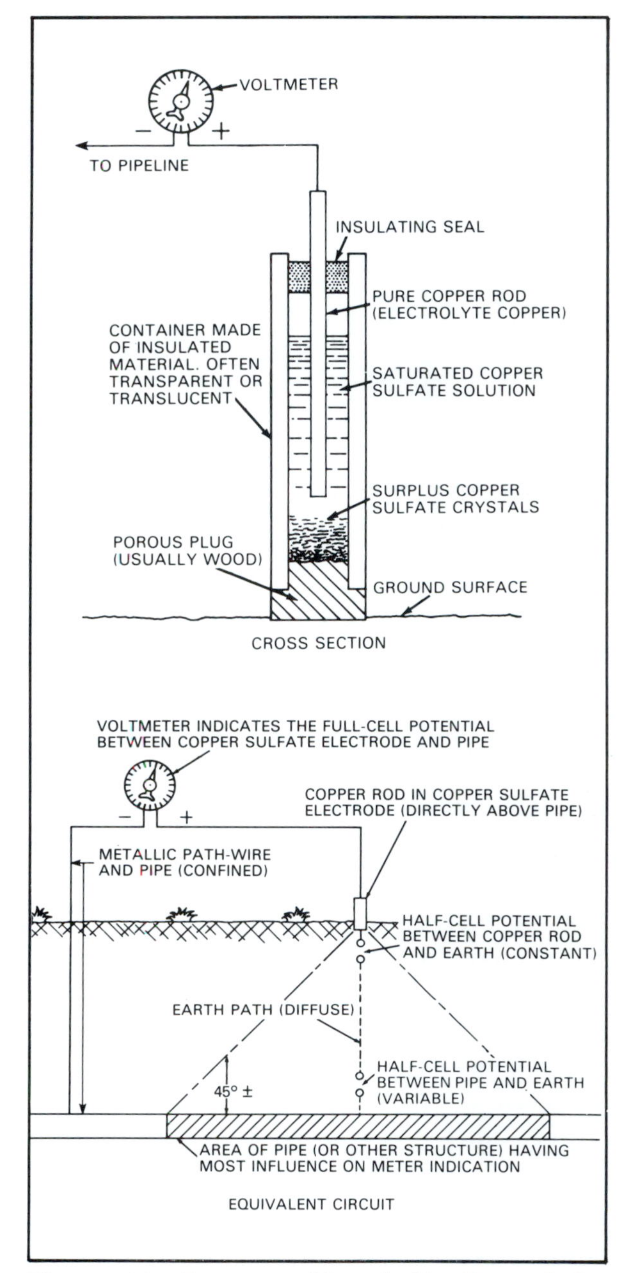

Figure 14. Copper sulfate half-cell used for measuring pipe-to-soil potentials

Other half-cells are sometimes used for contacting electrolytes, especially in laboratory and research work where they may be more suitable. These instruments are especially useful for locating cathodic and anodic areas of a structure and for monitoring the effectiveness of cathodic protection as a means of corrosion control. They can be used to indicate the direction and magnitude of current flow in an electrolyte.

TESTING FOR ANAEROBIC BACTERIAL CORROSION

A simple test may be used sometimes to check for anaerobic bacterial corrosion or, if needed, for sulfide corrosion from other sources. Sulfate-reducing bacteria cause production of hydrogen sulfide, which tends to form metal sulfides and corrodes metals. These corrosion products, when treated with hydrochloric acid, produce hydrogen sulfide, which can be detected by smell. Such tests should be made immediately after the samples are collected, because sulfides exposed to air oxidize readily into sulfates, to which testing is not applicable.

Rapid oxidation of sulfides can be a safety hazard. When air is admitted into a vessel that contains great quantities of iron sulfide, the rise in temperature from oxidation can ignite the hydrocarbons present, and fires or explosions may result. Piles of scale removed from oil storage tanks have been ignited by such *spontaneous combustion.* Safety procedures, important in all production operations, are critical in instances where such combustion can occur.

RADIOGRAPHIC EXAMINATIONS

When surfaces are not readily accessible for examination by other means, *radiographs,* using X rays or radioactive isotopes, can be useful in evaluating metal loss. Radiographic pictures provide accurate information on metal condition (fig. 15).

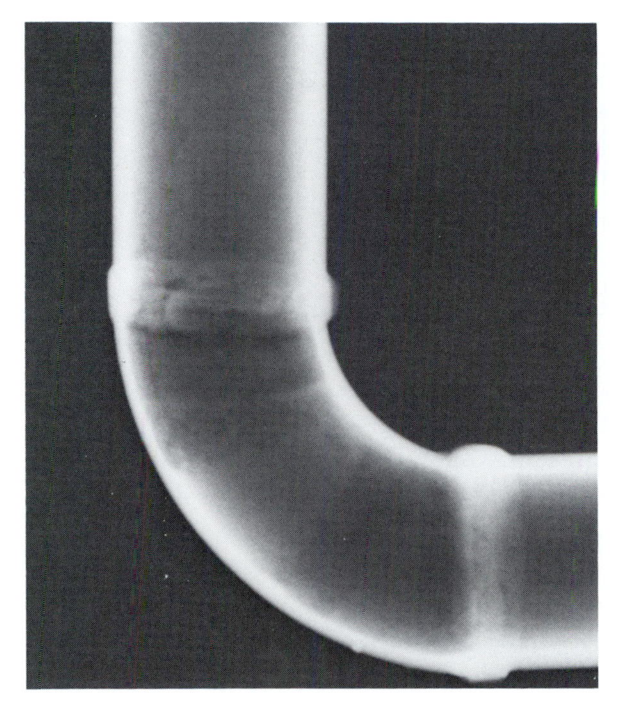

Figure 15. Radiographic view of corroded pipe

Periodic radiographic examinations may be justified to prevent equipment failure that would be extremely costly or disastrous. Such examinations can also be used to confirm the results of other tests. The main advantage of radiography is that it provides a picture record that can be understood by nontechnical personnel. The pictures may suggest the need for more effective control measures or provide assurance that a pictured article is serviceable.

ULTRASONIC DEVICES

Ultrasonic devices provide visual or recorded information on metal thickness and surface condition. These devices use ultrasonic energy generated by transducers that transform an applied high-energy frequency signal into high-frequency

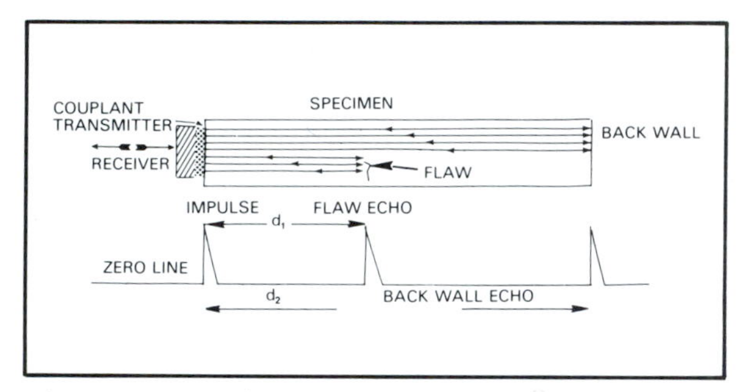

Figure 16. Pulse-echo technique of ultrasonic wave travel

mechanical energy. A sound wave generated by a transducer is transmitted by a liquid couplant to the metal wall. Ultrasonic waves travel through the wall until they reach an *interface,* or discontinuity. Usually this discontinuity is the opposite wall of the specimen being tested. The reflected sound wave is received and transformed into an electrical impulse by the transducer (fig. 16). The device measures the time between the impulse and its reflection, and then the thickness of the specimen can be calibrated, read, and recorded. This method, called the *pulse-echo technique,* is commonly used to measure metal thicknesses and to detect flaws.

Inspections should be made on new equipment as it is installed, and records should be kept for reference. Subsequent tests on the equipment, when compared with the original data, show the progress of corrosion. One drawback of this method is that tested surfaces must be free of scale and other substances, including liquids. To obtain reproducible results, the detector probe must be oriented carefully.

Trained operators are required to interpret the data. Like other tests requiring specialized equipment and procedures, these procedures are often handled by service organizations specially equipped and trained to provide the service.

ELECTRONIC PIPELINE INSPECTIONS

Some inspection devices can be pumped through a pipeline to make a continuous record of its wall condition. The liquid or gas normally pumped through the pipeline propels the device, which utilizes an electromagnetic flux leakage technique to detect both internal and external defects (fig. 17).

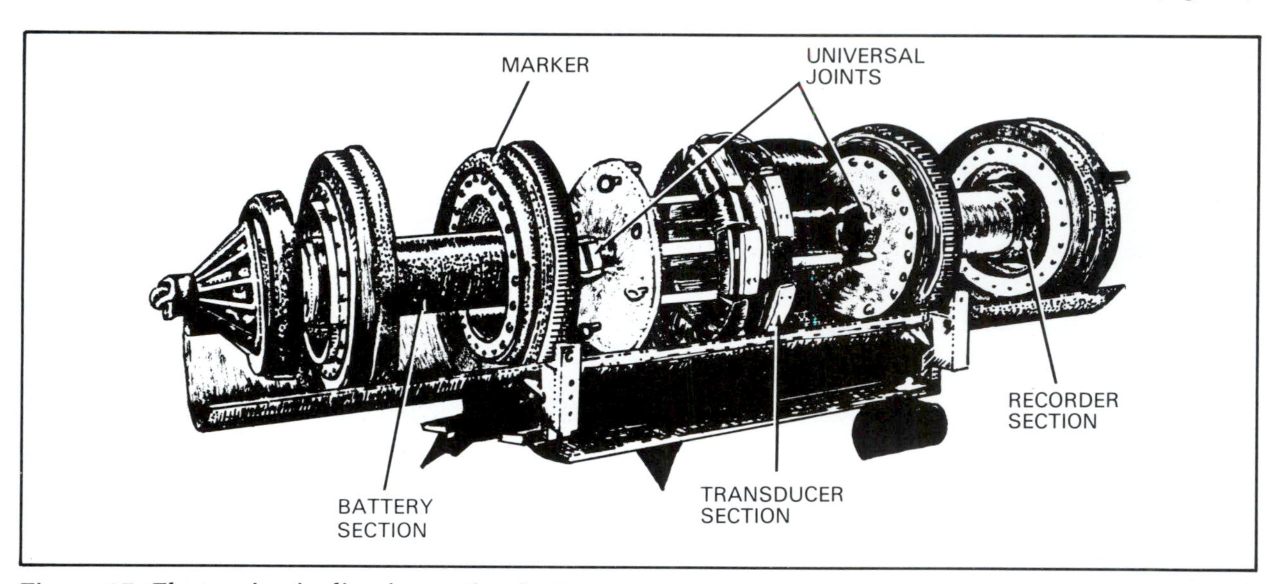

Figure 17. Electronic pipeline inspection instrument

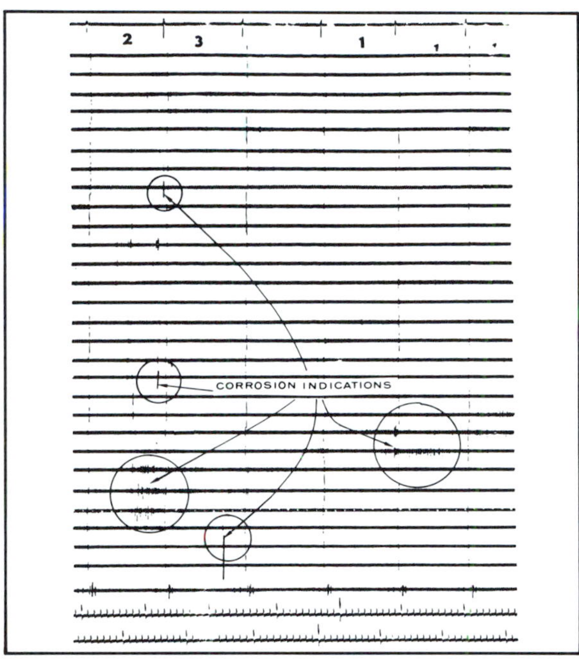

Figure 18. Record of electronic pipeline inspection

These defects are then recorded magnetically (fig. 18). Many production pipelines are too small to justify the use of such tools, but for the larger offshore pipelines where pipe failure can be disastrous, electronic inspection methods are feasible.

ELECTROMAGNETIC EXAMINATION OF CASING

An electromagnetic inspection tool that identifies casing damage is available. This tool shows damage to be inside, outside,

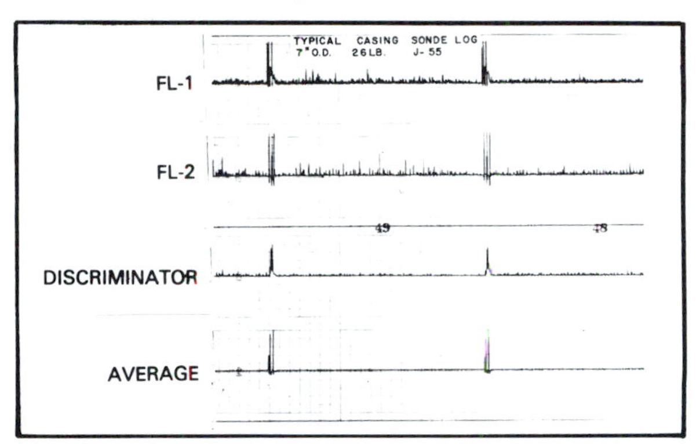

Figure 19. Inspection record of electromagnetic tool

isolated, or circumferential and produces a record of the findings (fig. 19). There are four channels that make up the device, two of which are flux-leakage channels. A third discriminator channel provides a record when an anomaly is inside the casing. The fourth channel records an average and makes it possible to determine whether or not the anomaly is circumferential.

Interpreting electromagnetic logs requires definite information on the well, as well as casing and tubing specifications. The logs may show the type of damage to each joint, the percentage of penetration, and other anomalies. Comparing successive logs can show the rate of corrosion. The accuracy of the well data required to interpret the logs explains the need for detailed records of well construction and other pertinent facts in the well's history. Unfortunately, design plans and construction specifications do not always provide accurate information about actual structures.

CALIPER SURVEYS

Calipers have been used downhole for surveying the insides of casing and tubing for many years. They can be used to locate and measure mechanical damage such as rod wear in pumping-well tubing strings.

Calipers are fitted with feelers that are held by spring pressure to the tubing or casing walls. Depending upon pipe size, tubing calipers may have from fifteen to forty-four feelers; casing calipers, from forty to sixty-four feelers (fig. 20). As the feelers extend into a corrosion pit or rod score, they mechanically change the setting on a potentiometer above the feeler assembly or record the deflection on a metal chart. Potentiometer data are transmitted through a conductor cable to an instrument on the surface, where a continuous log is recorded

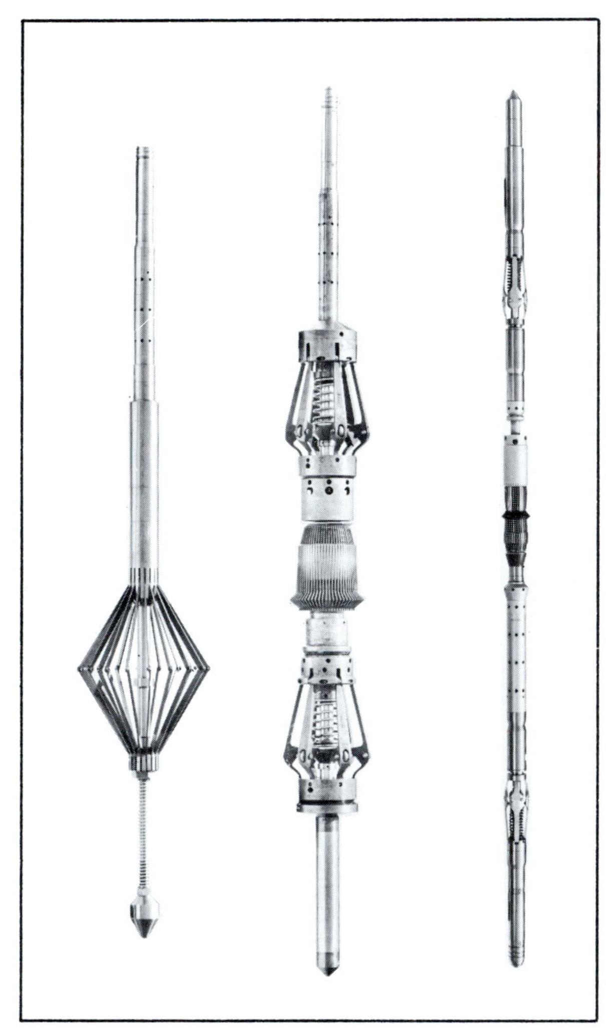

Figure 20. Casing and tubing profile calipers

on a chart (fig. 21). Maximum penetrations of pits and rod scores are printed out as numerical percentages, in addition to the penned traces of meter readings. Caliper surveys are particularly valuable in checking the effectiveness of inhibitor programs, if an initial survey is available for comparison.

Calipers generally enable operators to measure corrosion and wear of pipe in place, but they have some limitations. Scale or hard corrosion products may fill pits and prevent the feelers from entering; acidizing can help obtain more accurate surveys. The spacing of the feelers may result in some pits being missed. Plastic tubing linings can be damaged. Protective scales or inhibitor films can be removed or scored by the caliper feelers or by the wirelines on which the calipers are suspended.

MEASURING ELECTRIC CURRENT IN CASING

Outside corrosion of casing pipe is more severe in some areas than others, as a result of formation differences. There is a net discharge of current to the soil in corroding areas and a net collection of current in the other parts. The resulting electrical currents flow in the casing toward the areas of current discharge. The direction and magnitude of these currents can be determined by measuring the *IR* drop in the casing at a series of locations (fig. 22).

In the drawing, the first curve (*1*) indicates natural corrosion. The second curve (*2*) indicates corrosion induced by cathodic protection, a means of control that utilizes the corrosion process itself. Values to the left of the zero line indicate downward flow of current; values to the right, upward flow. From point *A* to point *B* on the first curve, more current is being discharged than is being collected, and the area is predominantly anodic. Increasing values in the direction of current flow show a net inflow of current from the soil, and decreasing values show a net outflow of current. Local cell action has no effect on the curves.

In order to calculate current from the $E = IR$ equation, casing size and weight and the electrical resistivity of the steel (corrected for temperature) must be known. Also, the casing must be clean and dry, because moisture can cause galvanic potentials greater than those being measured.

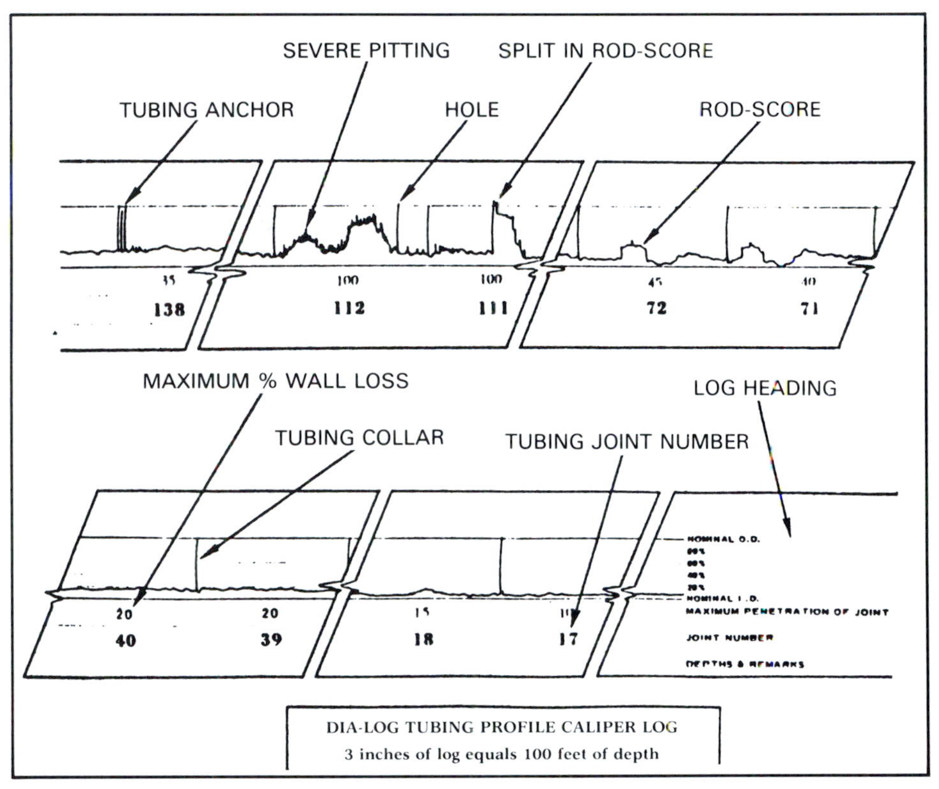

Figure 21. Tubing profile caliper log (Courtesy Dia-log)

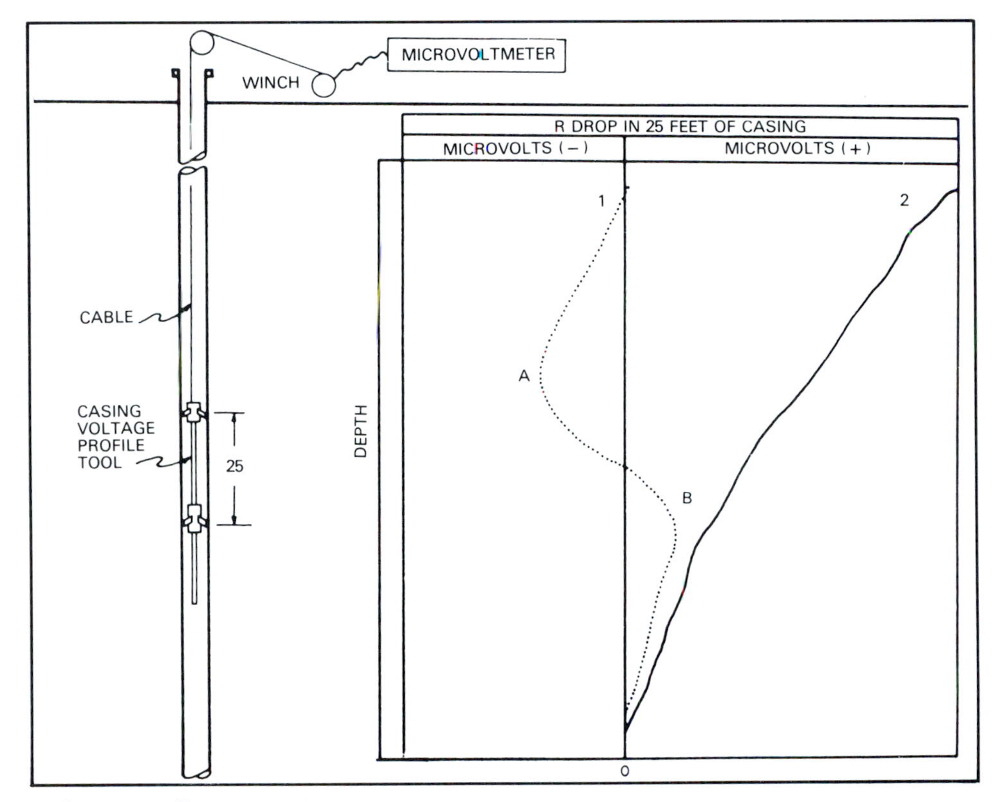

Figure 22. Measurement of current flow in casing pipe

This type of survey can measure natural currents, cathodic-protection currents, and interference currents.

INSPECTION RECORDS

In fields that have been producing for a number of years, a relatively effective and inexpensive method of estimating corrosion is the keeping of good records of equipment inspections. Operators are usually willing to exchange recorded information about any changes, failures, accidents, and abandonments that such *use tests* may indicate.

MEASUREMENT COMBINATIONS

Some instruments are devised to utilize more than one measuring technique to provide complete pictures of facilities under test. Sometimes more than one survey using different techniques may be needed for satisfactory evaluation of equipment condition.

Self-Test

3: Detection and Measurement

(Multiply each correct anwer by four to arrive at your percentage of competency.)

Fill in the Blank

1. The __________ ____________ in production fluids can suggest the extent of corrosion and the effectiveness of control measures.

2. Instantaneous corrosion rates can be measured by means of an electrochemical phenomenon called ______________ _________________________.

3. A _________________ ____________ ____________ measures hydrogen absorbed into steel as a result of hydrogen sulfide corrosion.

4. _________________ __________________________ causes steel that contains atomic hydrogen to rupture from relatively small stresses.

5. Nonpolarizing electrodes, or _____________________, are used to determine whether a metal is cathodic or anodic to an electrolyte.

6. Rapid oxidation of sulfides can cause ____ ____________________, which can be a safety hazard.

7. _________________________ _______________ can be used when metal surfaces are not accessible to other forms of examination.

8. The _____________________ technique measures metal thickness by recording the progress of sound waves.

9. Electromagnetic casing inspections can show whether corrosion damage is (1)______________, (2)________________, (3)_________________, or (4)_______________________________.

10. _________________ utilize feelers which survey the insides of pipe.

True or False

11. _____ Corrosion rate is an important factor in determining corrosion control measures.

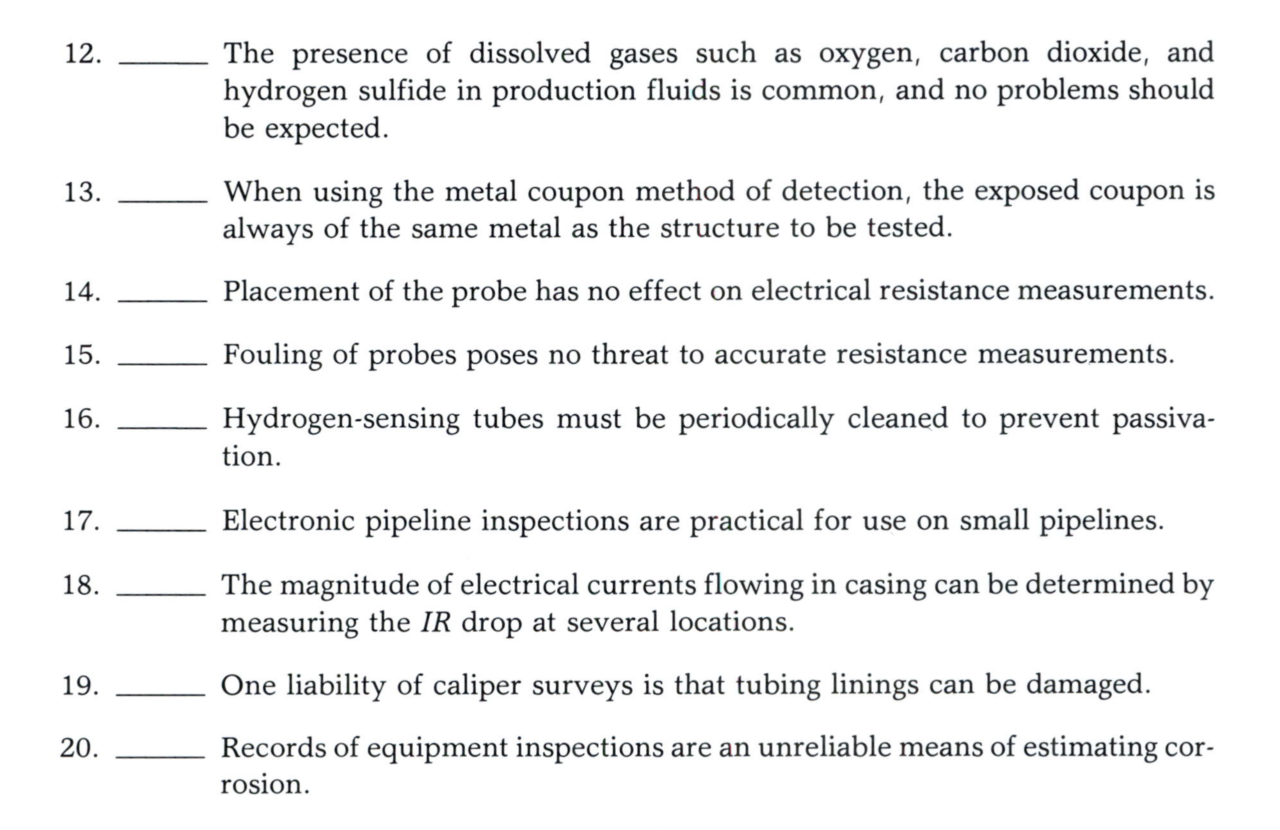

12. ______ The presence of dissolved gases such as oxygen, carbon dioxide, and hydrogen sulfide in production fluids is common, and no problems should be expected.

13. ______ When using the metal coupon method of detection, the exposed coupon is always of the same metal as the structure to be tested.

14. ______ Placement of the probe has no effect on electrical resistance measurements.

15. ______ Fouling of probes poses no threat to accurate resistance measurements.

16. ______ Hydrogen-sensing tubes must be periodically cleaned to prevent passivation.

17. ______ Electronic pipeline inspections are practical for use on small pipelines.

18. ______ The magnitude of electrical currents flowing in casing can be determined by measuring the *IR* drop at several locations.

19. ______ One liability of caliper surveys is that tubing linings can be damaged.

20. ______ Records of equipment inspections are an unreliable means of estimating corrosion.

4

Methods of Corrosion Control

OBJECTIVES

Upon completion of this section, the student will be able to:

1. Describe six commonly used methods of corrosion control.
2. Define *alloy*.
3. Explain why different combinations of metals are used in production structures.
4. Give four reasons why coatings deteriorate.
5. Explain the distinction between *soluble* and *dispersible* inhibitors.
6. Give some advantages and disadvantages of various types of inhibitor treatments.
7. Give three important considerations in selecting an inhibitor.
8. Explain the importance of prohibiting oxygen from entering production systems.

CONTROL MEASURES

Several measures can be taken to limit corrosion damage. First, production systems should be designed so that their liability to corrosion is minimal. Construction materials, whether pure metals, alloys, bimetallic combinations, or nonmetals, should be chosen for their corrosion-preventive properties. In spite of these structural precautions, the use of insulation, inhibitors, or measures to eliminate corrodants may still be necessary to interrupt corrosion cell activity. Cathodic protection, a complex corrosion-mitigation process, will be covered in the next chapter.

SYSTEM DESIGN

Corrosion control begins with system design. The properties of produced fluids and the rates at which they flow through piping are important considerations in planning a production system.

Flow rate may be a difficult factor to evaluate in the design stage. Rapid, turbulent flow of oil in pipes can prevent brine damage, keeping the brine dispersed and preventing it from separating out. In such cases, piping should be sized so that oil flow is turbulent and as nearly continuous as is practical. Brine can separate from oil that has been treated for pipeline transport if the oil is allowed to stand or is pumped so that flow is sinuous or nonturbulent. Likewise, when a limited quantity of corrosive oil must be handled, it is better to pump it periodically at high rates than to keep it moving slowly for longer periods. This measure will at least sweep the system, and only the brine in the line fill will settle during idle periods.

When corrosive oil is to be handled, piping should be arranged to avoid dead ends where scale and brine can collect and remain undisturbed. Such dead ends are frequently at manifolds where flanged tees are provided for ready extensions or new connections. Scraper traps should be of the flow-through variety so that they stay clean.

CORROSION-RESISTANT CONSTRUCTION MATERIALS

Construction materials should be selected for their resistance to corrosion under the expected working conditions (fig. 23). Carbon steel, with its various grades of strength, hardness, ductility, and notch toughness is the preferred construction material for most structures. However, its susceptibility to corrosion necessitates the use of other materials.

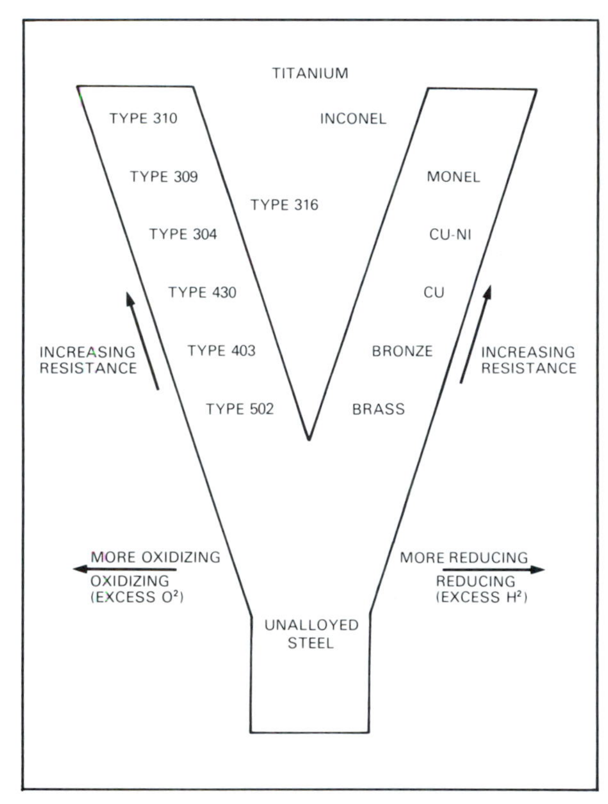

Figure 23. Corrosion resistance of various metals in oxidizing and reducing environments

Pure Metals

Some pure metals have good corrosion resistance. Aluminum, nickel, and copper are among such metals. Silver, gold, and platinum may be used for plating small surfaces such as electrical contacts in switch gear. However, silver and, to a lesser extent, copper combine readily with sulfur compounds, especially hydrogen sulfide; this tendency limits their use in operations where sulfur may be encountered.

Alloys

Alloys, solid solutions of two or more metals, are often used for production structures because of their castability, machinability, and corrosion resistance. (They are also more expensive than steel structures.) Alloys are divided into two classes, *ferrous* and *nonferrous.* Ferrous alloys contain iron as the major ingredient, with significant proportions of other elements such as chromium, nickel, manganese, and molybdenum. The stainless steels are also in this category. Technically, all steels could be classed as ferrous alloys since they contain small proportions of carbon (usually 0.2–1.0 percent). The carbon is in the form of iron carbide. Nonferrous alloys contain less than 50 percent iron, even though iron may be the largest single component. Brass, copper-nickel alloys, and the iron-nickel-chromium alloys, though nonferrous, may be considered extensions of the stainless steel series.

Alloys have many uses. Tube bundles of heat exchangers and condensers are made of alloys that have good heat-transfer characteristics and form corrosion products that are not likely to plug the tubes or restrict the flow of fluids around them. Brasses such as Admiralty Metal (70% copper, 29% zinc, 1% silicon) may be used for tubing.

Requirements other than corrosion resistance often dictate the use of nonferrous alloys. These alloys may be drawn into small, thin-walled tubes that are readily formed into various shapes and easily joined to tube sheets with liquid-tight joints.

Since tubing heads are made of thicker material that may be different from that in the tubes, the use of alloys with similar composition can avert the possibility of galvanic corrosion. As a rule, similar alloys do not suffer severe bimetallic corrosion. Liquid-tight joints formed by rolling the tube ends into holes in tube sheets can be opened up by small amounts of damage. For this reason, baffles that are used to direct the flow of fluids around tubes should be made of materials compatible with the tubes.

Bimetallic Combinations

Metals can be used in beneficial combinations, either to provide corrosion-resistant coverings or to deliberately set up corrosion cells that will prevent long-term damage.

A noble metal with good corrosion resistance can be used as a covering for a metal such as steel that is suitable for production because of its price or strength. A common example of this type of covering is the tin can, for which thin steel sheets are coated with extremely thin coverings of tin. It is possible to apply the tin without imperfections, and it is so resistant to corrosion in most atmospheres that the contents of cans may resist spoiling for many years. Once the tin coating is punctured or broken, corrosion accelerates.

Steel vessels lined with metals such as nickel are widely used in oil refineries and, to some extent, in oil-treatment facilities. In such vessels, the lining must be free of imperfections that permit corrosive electrolytes to contact the steel vessel. Because

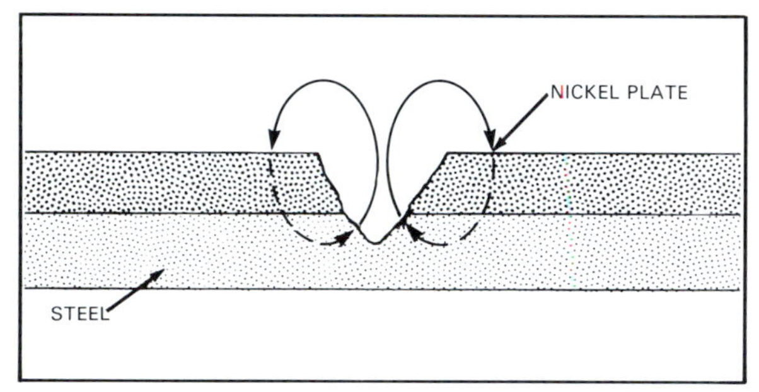

Figure 24. Effect of surface imperfection in cladding nobler metal on steel

the nickel offers a large cathode and the steel is anodic in a small area, a little current can result in rapid penetration (fig. 24). The small current is not enough to polarize the large cathode, so cathodic control does not act to slow the reaction. Extreme care should be taken in construction, handling, and operation of all vessels to avoid imperfections in their linings.

Zinc, which has good resistance to many corrodants, offers large areas as anodes, and the small area of exposed steel is cathodic and quickly polarized (fig. 25). Protection is not seriously hampered by small voids in the zinc coating. The preferred method of applying zinc is hot dipping. Zinc may also be applied by electrodeposition or painted on as a powder dispersed in inorganic paint vehicles.

In some cases it is practical to use different metals in a structure, despite their combined tendency to cause corrosion. Freshwater piping regularly consists of steel pipe with brass valves, but corrosion rates are low, and the anodic steel pipe offers such comparatively large areas and masses that serious damage is not likely. In salt water, where water conductivity and corrosion tendency is higher, rapid corrosion may occur inside pipe that is adjacent to brass or other alloy fittings. Such corrosion can be minimized by using tight-fitting, nonconductive inner sleeves a few inches long at pipe ends, so that cell currents must travel greater distances through the relatively high resistance of the liquid. This moderate lengthening of the electrolytic path when its cross-sectional area is small is sufficient to overcome the weak cell voltage.

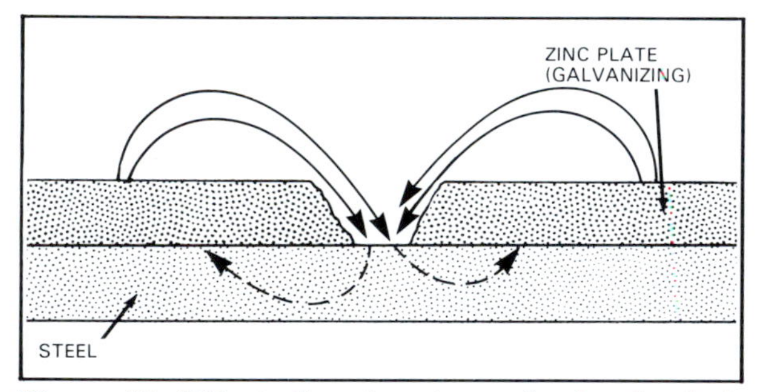

Figure 25. Effect of surface imperfection in cladding baser metal on steel

The use of small bits of metal that are anodic to the main body of a structure can have disastrous consequences. Using a small steel nipple in a copper piping system is one such mistake that happens too often. In major construction projects, using the wrong kind of bolts, rivets, or welding rods can prove to be a costly blunder.

Nonmetallic Materials

Nonmetallic materials such as asbestos cement, which has long been used for pipe, roofing, and siding, have seen some use in production structures. Even concrete, applied as Gunite linings in tanks or as pozzolanic linings for pipe, has been used in some situations where corrosion is severe. Fiberglass-reinforced pipe and pipe made with PVC, polyethylene, and polypropylene are important nonmetallics.

Plastics, especially those reinforced with glass fibers, are widely used, but those plastics suitable for use in extremes of temperature, weather, and sunlight tend to be expensive. Their greatest applications have been as components of protective coatings.

COATINGS AND LININGS

The earliest corrosion control measures consisted merely of brushing away the corrosion products from an article to make its surface more uniform. Then, a light covering of oil was added by hand. This procedure was followed by burnishing and treatment with various oils such as lanolin. More durable coatings made of drying oils such as linseed oil have been used for many years.

The use of paints and coatings is intended to break up corrosion cells by isolating the metal, physically or electrically, from the electrolyte. This method is still the most widely used corrosion control measure. During the last few decades, paints, coatings, metal surface preparations, and application procedures have been improved so that this field has developed into an exact science. Since factors such as weather can affect the manufacture, marketing, storage, application, and exposure of coated surfaces, the effects of coatings are less predictable than might be desired. The use of coatings is also complicated by the fact that they are used for appearance, sanitation, safety, identification, and advertising. The National Association of Corrosion Engineers (NACE) and other societies have developed standards and recommended practices to maintain this method of corrosion control.

The properties required of coatings can be exacting. Most coatings have thermal expansion rates different from the metals to which they are applied. The attachment, or *bond,* between coating and metal needs to be strong enough to resist the strain of temperature changes. Most coatings are *thermoplastic:* they lose many or all of their protective properties at elevated temperatures. Those coatings that set or solidify by *polymerization,* the formation of larger molecular units of structure identical to their original form, may be stable up to their decomposition temperatures. The newer, extremely tough coatings with thicknesses less than 10 mils achieve bonding by forming themselves around an *anchor pattern*—minute projections from the metal surface produced by sandblasting, shot blasting, or chemical etching. Anchor patterns must not have sharp projections higher than the thickness of their coatings.

Types of Coatings

A variety of coatings has been developed to serve various corrosion-prone situations. A unique and quite useful coating is inorganic zinc silicate. It is applied as a thin, paintlike coating. It can be set by heat, but a sprayed-on curing agent that causes solidification of the silicate is most often used. The zinc, in the form of fine powder, provides an anodic surface to protect the underlying steel at small breaks in the coating. Zinc silicate may be overcoated with vinyl or other coatings to conform with color schemes and prolong its service life. It has found wide use in coastal atmospheres, tanker compartments carrying sweet crude oil and seawater ballast, tank roofs, and areas where rainwater drainage is poor.

Among the many other coating materials available are phenolics, epoxies, and polyesters. Thermosetting plastics like the phenolics are widely used in measurement equipment such as positive-displacement meter provers. Automatic dump tanks, which measure oil by counting the times a tank with a carefully calibrated volume is filled and emptied, also require linings that are resistant not only to corrosion but to paraffin deposition as well.

Some thermosetting, catalytic-setting, and polymerization-setting coatings are so tough and moisture resistant that films 10–40 mils thick can provide good protection to buried pipelines. Thermoplastic pipe coatings can be applied hot in the liquid state, taking the form of seamless sheaths. For buried pipelines and other structures, rolls of plastic sheeting with a preapplied adhesive, or *tape wrapping,* are also widely used. It should be remembered that coatings suitable for use underground or underwater may not be suitable for exposure to sunlight and weather.

Application of Coatings

Often, the most feasible (or available) time to apply coatings is during construction, so long and effective service life is a must. The inconvenience of recoating well casings, large pipelines, and offshore structures justifies the need for initial coatings that will last.

NACE and other agencies prescribe suitable surface preparations for different coatings. Metal surfaces must be free of dust, oils, corrosion products, or any foreign materials that may adversely affect the coating while it is in the liquid application phase. Surfaces *must* be moisture-free; even invisible films of condensed moisture cannot be tolerated by most coatings. Thicker coatings tend to be more elastic and can absorb some strains without disbonding. Fortunately, concrete and steel have about the same coefficients of thermal expansion, so satisfactory bonds can be achieved.

After a corroded steel surface has been cleaned and prepared, a layer of low pressure resistant, fiberglass-reinforced plastic can be applied. Such coatings may take the form of glass fiber mats embedded in the freshly applied plastic. Guns that spray plastic and chopped fiber simultaneously are also available. Polymerization of plastics may depend upon the release of solvents, action of catalysts, mixtures of two components, or application of heat. Plastics not containing solvents are referred to as *100 percent solids,* since all components are polymerized to the solid state.

Thermosetting coatings usually require several applications. After each application, a vessel may be baked at an intermediate temperature of about 250°F to drive out solvents, with a final bake of 400°F–450°F to polymerize all into a homogeneous coating. The total thickness of such coatings is often in the 6–10 mil range.

Many plants are in the business of applying coatings to new as well as used pipe. Surface preparation is exacting. All foreign materials, especially oil and grease, must be removed, and low-profile anchor patterns must be developed for the thin coatings. Metal temperatures at the time of the coating application should always be above the atmospheric dew point to avoid condensation. Plant locations are selected to minimize the costs of transporting pipe to and from plants.

Deterioration of Coatings

Coatings may deteriorate for surprising reasons. The first efforts to protect buried pipelines by coating them with molten bitumens such as asphalt and coal tar failed because changes in soil volume and moisture caused uneven pressures that soon penetrated the coatings. This action is called *soil stress.* This problem can be remedied with shields of asbestos felt, which distribute the pressures more evenly; also, reinforcement with embedded glass fibers reduces the plasticity of the bitumens.

During storage, transfer, and construction, coated and lined articles such as pipe require special handling to avoid damage. Construction cost savings achieved by using damaged pipe usually amount to a bad bargain. Lined pipe, especially that lined with concrete, should not be thrown or dropped from trucks or allowed to bump roughly into other objects. Lifting hooks that fit into pipe ends should be used cautiously. Other pipe-handling equipment—slings, cradles, cables, and chains—should be designed to prevent damage to coatings.

Coatings exposed to seawater may be destroyed by mollusks such as barnacles and oysters. The growth of these mollusks after their attachment to coated objects causes pressure which displaces and penetrates coatings, whether the coating is hard and thick like asphalt mastic, or soft like paraffin or heavy grease. Such deterioration may be remedied by burying structures under the seafloor or encasing them in concrete.

INSULATION

Sometimes, breaking the metallic path for current flow in corrosion cells is a good control measure. Insulating flanges and unions can be used to interrupt current flow. The tendency of above-ground facilities to serve as cathodic current gathering areas for the outsides of well casings can be remedied by insulating flow lines and other well connections. Possible bypasses of such insulation (e.g., metallic electrical conduits) should not be overlooked. Electrical insulation can also be used to interrupt the flow of stray currents, to control the distribution of cathodic protection currents, and occasionally to electrically isolate metals that would cause galvanic corrosion.

INHIBITORS

Inhibitors are substances that slow or prevent chemical reactions. Inhibitors are widely used as additives to acids, cooling water, steam, brine, oil-brine mixtures, gases, and refined products such as gasoline and lubricating oil. In many situations, such as downhole in oil and gas wells, inhibitors may be the only applicable remedy for corrosion. The quantities of the agents used are not sufficient to neutralize the corrodants; instead, these agents form films on the metal surfaces to protect them from the electrolytes. *Passivating agents,* which make metal surfaces passive rather than anodically and cathodically active, can also reduce corrosion rates.

Most inhibitors depend upon their molecular structure for effectiveness. Strongly active cations or anions are joined to massive organic ions, which are chemically inactive or sluggish. Depending on the makeup of these molecules, the active parts are attracted to either the cathodic or the anodic metal surfaces. Often a monomolecular film is strongly bonded to the metal by a process called *chemisorption.* The film thickness may be increased by additional, less strongly bonded layers. Some films may be thick enough to be visible, often appearing oily and water repellent. They may be *lipophilic,* attracted to oily substances, so that the metal surfaces to which the films are applied are wetted by oil rather than water. Oily films also have high electrical resistance, so they interfere with two of the elements of corrosion cells. The films separate metal surfaces from electrolytes and interrupt electrical circuits.

Oil Field Inhibitors

Oil field inhibitors are usually furnished in concentrated form to improve stability and to reduce packaging and transportation costs. For convenience in proportioning and injecting, many inhibitors are diluted in such solvents as crude oil or kerosine before they are used.

Most oil field inhibitors are liquids that may be classified as being *soluble* or *dispersible* in either oil or water. When an inhibitor-solvent mixture remains clear, the inhibitor is usually considered to be soluble in that solvent. A dispersible inhibitor can be evenly dispersed in another liquid by moderately shaking the mixture. Dispersions that break in less than one minute are said to be *temporary.* These temporary dispersions are most useful when injected into oil streams in turbulent flow so that dispersion can be maintained. Rapid breaking can facilitate the formation of inhibitor films in some cases. More stable dispersions that maintain uniform proportions are used for more extensive systems and require fewer applications. Inhibitors with combinations of properties can be formulated so that they are partly soluble and partly dispersible.

These thin, fragile inhibitor films can be subject to any one of a number of weakening factors: high temperatures, rapid flow of well fluids, brine, sand, or the rubbing action of sucker rods and wirelines. Therefore the durability, or *persistence,* of inhibitor films is an important quality. With regard to cost and convenience, films that remain effective for several days or weeks are preferable to those that require continuous injection. Inhibited electrolytes that remain quiet and are sealed to exclude oxygen – casing packer fluids, for instance – require inhibitors that can last without replenishment for years.

Weighted Inhibitors

Most inhibitors are liquids with specific gravities in the 7–12 pounds per gallon (ppg) range. By using various solvents, gravities can be adjusted so that they are heavier than those of oil or brine. These weighted liquid inhibitors, in capsule form, are less expensive to apply than some other inhibitors. They fall to the well bottom through the fluids being produced and then gradually disperse into the fluids. Weighted encapsulated inhibitors are designed to be unaffected by oil or dry gas. Upon contact with water, however, they break up, releasing the inhibitor.

There are two classes of such encapsulated inhibitors, based on the rate at which the inhibitor is released. *Fast-release* inhibitors are released immediately upon exposure to water. *Slow-release* inhibitors sink through the bottomhole water, yielding the inhibitor slowly. Release rates are increased by higher temperatures and concentrations of salts in brines. Fast-release inhibitors need a storage space such as a rathole at the well bottom. An advantage of this method is low treating frequency. Fast-release treatments can be used to establish protective films, followed by slow-release treatments to maintain them.

The choice of a weighting agent for a liquid weighted inhibitor is important. Zinc chloride, for instance, reacts with hydrogen sulfide to form zinc sulfide, which is an emulsion stabilizer. Emulsions that form are difficult or impossible to break. Wells producing hydrogen sulfide require inhibitors weighted with heavy organic chemicals. The weighted inhibitor method has certain favorable factors such as simplicity and ease of application, but possible side effects and other limitations require that caution and good judgment be exercised at all times.

Inhibitors can also be formed into heavy sticks. They can be formulated to dissolve at various temperatures and at differing rates so that a *time-release* effect is obtained.

Selecting Inhibitors

Several properties need to be considered in choosing an inhibitor. One is compatibility with other chemicals. Usually the small proportions of inhibitor (measured in parts per million) cause no problems, but some chemicals do react with each other, creating unexpected effects. The end results of mixing chemicals should always be investigated beforehand.

Another consideration is the emulsion-forming tendency of an inhibitor. Tests should be made to determine whether the use of a proposed inhibitor would result in the formation of stable emulsions.

Elevated temperatures may affect inhibitors. This effect is influenced by pressure, the presence of water, and other conditions. Bottomhole temperatures may range up to 600°F (315°C) or higher, and inhibitors stable at these temperatures are available.

Downhole conditions may not always be fully understood. Tests can be made using an inhibitor in varying proportions to the production fluids to indicate the inhibitor's effectiveness. By applying chemical principles, the results of preliminary tests, and the advice of experts, an inhibitor can be selected. Sometimes a mixture of two or more inhibitors proves to be more effective than the combination would suggest, because of the synergistic effect by which one material affects the performance of another.

Methods of Inhibitor Application

The type of equipment in a well influences the type of treatment to be used. For example, wells with tubing strings set on packers do not have circulation between the tubing and the tubing-casing annulus, and inhibitor injection is difficult to accomplish. Depending on the method of recovery – pumps, artificial gas lift, or formation pressure – corrosion control requirements vary. The three basic forms of introducing inhibitors into formations are continuous treatments, squeeze treatments, and batch treatments.

Continuous treatment. The advantage of continuous treatment is that the concentration of an inhibitor in the treated fluids can be maintained at a constant level and changed as needed. Concentrations may range from a few parts per million to fifty or more. Treatment of pumping wells is comparatively easy. The inhibitor can be injected (usually under power) with a small side stream of the produced liquids into the annulus. Concentric tubing strings or macaroni strings can also be used. A pressure-actuated valve can be installed near the bottom of a tubing string above a packer to admit a mixture of inhibitor and solvent, which is pumped into the annulus at the surface.

Squeeze treatment. The squeeze method, whereby substances are introduced into producing formations as they are needed, can be used for applying inhibitors. Using the formation as a reservoir for the inhibitor, squeeze treating ensures that only treated fluids will contact downhole equipment. An inhibitor dispersed into the formation should be gradually released into the produced fluids over a period of from three to eighteen months, depending on the inhibitor, the way it is applied, and formation conditions. The squeeze method can be used in wells producing from multiple pays.

In squeeze treatments the proportion of inhibitor to diluent may vary widely. Commonly used amounts are from one to five 55-gallon (gal) drums of inhibitor and twenty-five to one hundred drums of diluent. Additional fluid is added for the tubing fill, plus enough to force the mixture the desired distance into the formation. This procedure provides a relatively high concentration of inhibitor both during the application and for a short time after production resumes. Good protective films should be laid down during these periods, and, given good persistence characteristics, they will be maintained by the lower but steady amounts in the fluids produced later.

Batch treatments. This type treatment is generally used in conventionally pumped wells. The treating material is dumped down the annulus, followed by a flush volume. This process is repeated at intervals of one or two weeks.

Effects of Treating Formations

Treating formations can be a complicated process. Placing inhibitors in producing formations results in an adsorption-desorption process that makes the method effective. Inhibitor is adsorbed onto the surfaces of formation particles or on the insides of passageways in hard formations. Then, most of the inhibitor is slowly desorbed as the fluids flow into the wells. Sand and limestone give up adsorbed substances readily, but clays, which adsorb large quantities, retain large portions. That proportion desorbed from the clays may provide adequate protection and make the process feasible.

The effects of a treatment on producing formations should be known beforehand. Permeability, that vital property, should not be reduced significantly. Sands are not usually affected seriously, but clays may swell when alien fluids are introduced, tending to plug flow channels. Tests described in the product literature should show swelling tendencies of clays in core samples.

The emulsion-forming tendencies of proposed inhibitors with formation fluids should also be predetermined by tests. Stable, viscous emulsions can effectively plug formations, requiring expensive retreatments to restore productivity. Inhibitors or diluents may react with substances encountered in the formations, forming solid or unctuous precipitates and reducing productive capacity.

Gas wells with low formation porosity can be plugged by liquid injections. In such cases, an inhibitor must be atomized into a gas such as nitrogen and then forced into the formation under additional gas pressure.

Water Injection Treatments

Oil field brines are often reinjected into underground formations for improving oil production or for disposing of the brines. Preventing oxygen from entering the brines during separation, collection, transfer, storage, treating, and injection is a good check on corrosion. Inhibitors are needed in most cases, however. They may be water-soluble or water-dispersible. Inhibitors should be prevented from plugging formations, which would be indicated by increases in the pressure required for injection.

Water supplies other than oil field brines can be used for formation flooding. Corrosion control measures such as deaeration and inhibitor use are often required for such waters.

Oil field brines and waters suitable for flooding are not pure substances. Bacterial slime or sediment can foul the surfaces of transfer piping, tanks, and other containers, preventing contact of inhibitors with metals. It may therefore be necessary to run scrapers through the pipes and clean the interiors of vessels before inhibitor treatments and periodically thereafter. Cleaned systems are usually treated by slugging with strong inhibitor mixtures, followed by continually injecting smaller amounts to maintain protection. Such cycles may need to be repeated from time to time.

Cooling Water Treatment

Cooling waters usually need treatment to prevent hardness, scale formation, algae growth, or bacterial slime formation. Since some of the cooling water may tend to evaporate, systems may become so concentrated as to be classed as brines. Inorganic passivating inhibitors such as the chromates may be used, but their toxicity introduces problems in disposing of wastewater. The pH values of such inhibitors should be carefully controlled to avoid low values favorable to acid-induced corrosion and high values favorable to alkali-induced scale formation.

OXYGEN REMOVAL

Oxygen, though rarely present in oil and gas formations, can be readily absorbed into oil, gas, water, and brine. Problems with oxygen occur in aerated systems, water disposal, waterfloods, and cooling systems. On steel surfaces in fresh water, oxygen usually causes an overall type of corrosion that may not require specific control measures; use of galvanized tanks and piping is usually advisable to provide longer life and reusability of the materials. In salt water or brine, oxygen causes pitting, which can lead to early penetrations and failures.

The most obvious way to prevent oxygen-related corrosion is to prevent oxygen from entering a production system. This can be accomplished with careful attention to structural materials and production operations. When oxygen is in the system, chemical means such as scavenging, vacuum deaeration, and countercurrent stripping may be used.

Preventing Oxygen Entry

Oxygen can enter production streams in many ways. Air can be drawn into the annulus of a pumping well through a faulty connection, such as the annulus valve or seal. Air can also enter at the polished rod packing. Careful attention must be paid to the details of the annulus connections and the polished rod packing to eliminate this source of entry. The annulus of a pumping well should be kept filled with natural gas under slight pressure by tying the casing into the flow line or other source of gas. Air can be introduced into a gas-lift well along with gas. Using air-free gas or removing the oxygen from the gas can solve this problem.

Centrifugal pumps may suck in air through faulty packings around their shafts. A pump that leaks water when idle will probably draw air when it is running, so the packing should be replaced. Back-pressure of water may shut off the packing when the pump is idle, and this possibility should be considered if other points of air entry cannot be found.

Air above liquids in storage tanks and other vessels can also be a source of oxygen entry. Such air spaces must be filled with air-free gas at pressures of at least 3–4 ounces (oz) to avoid drawing in air when the liquids are pumped out. Gas should be supplied at rates that will maintain pressures greater than atmospheric pressure at all times. As a rule of thumb, a 500-barrel (bbl) tank requires at least a ¾-inch (in) gas line; a 1,000-bbl tank needs at least a 1-inch line. Smaller lines with higher pressure can be used, provided that pressure regulators to reduce pressure are installed at or near the tanks. Most of the tanks involved are not pressure vessels, and internal pressure is limited to the equivalent of the weight of the tank roofs. Layers of oil have been used over brine in tanks to prevent oxygen from entering into the brine. However, the oxygen, which is more soluble in most crude oils than in brines, is readily transferred through the oil-water interface into the brine, so the method is not effective. Due to the venturi effect, it is possible for air to enter at small leaks, particularly at flanges. Such points of entry are difficult to find, but they should be sought if other sources have been eliminated. Careful inspection with portable oxygen meters can often locate the leaks.

Protective coatings have been successful for preventing oxygen entry. Piping lined with cement (usually the pozzolanic type) is often used for handling aerated brine. The cement is applied centrifugally by spinning the pipe to obtain a dense, strong lining. Plastic coatings up to 0.018 inch thick and interliners offer a potential for better performance. Frp pipe can also be used for handling aerated brine, but its use is limited because of its high cost.

Chemical Removal

Sometimes it may seem impossible to eliminate all sources of oxygen entry, particularly when aerated water is the only available means of flooding. There are certain chemical and mechanical methods of neutralizing aerated water, but they are generally quite costly. Oxygen that is unavoidably present may be removed by scavenging, vacuum deaeration, or countercurrent stripping.

Scavenging is a means of oxygen removal that relies on chemical reactions with substances that are introduced into the system. Some commonly used scavenging agents and their approximate costs are as follows:

1. Sodium sulfite, 2 mills/bbl of brine (1 mill = 0.01¢);
2. Hydrazine, 4–5 mills/bbl;
3. Liquid sulfur dioxide, 1 mill/bbl (for large installations);
4. Sulfur dioxide from a sulfur burner, 0.3–0.4 mills/bbl (after initial capital investment).

Vacuum deaeration, followed by scavenging, costs about 0.12 mill/bbl to reduce oxygen to 0.5 ppm. Scavenging to remove the remaining oxygen would increase the cost to about 0.3 mill/bbl. This method requires a high capital investment.

Countercurrent stripping with natural or inert gas costs about 0.1 mill/bbl to reduce oxygen to 0.05 ppm and also requires a high capital investment.

Self-Test

4: Methods of Corrosion Control

(Multiply each correct answer by three to arrive at your percentage of competency.)

Fill in the Blank

1. Ideally, corrosion control begins with _______________ _______________.
2. High, turbulent flow of oil in pipes can prevent _______________ _______________.
3. An __________ is a solution of two or more metals.
4. Insulation is used to ____
   ____.
5. The bond between a coating and the protected surface should be strong enough to resist
   ____.
6. Coatings may deteriorate as a result of (1) ______________________________
   (2) ____________________ ,(3) ____________________, or (4) ____________________.
7. ______________________ are substances that slow or prevent chemical reactions.
8. ______________________ inhibitor films attract oil rather than water.
9. Most oil field inhibitors are liquids that may be classified as being ________________ or
   ________________________ in either oil or water.
10. Durability, or ______________________, of inhibitor films is an important quality.
11. Three things to be considered in selecting an inhibitor are:
    (1) ____,
    (2) ____, and
    (3) ____.
12. The advantage of continuous treatment is that ______________________________
    ____.
13. To ensure that only treated fluids will contact downhole equipment, the _____________ method should be utilized.

14. If inhibitors are used in water-injection systems, increases in the pressure required for injection might indicate that ______________________________.

15. In salt water or brine, oxygen usually causes ______________, which can lead to early penetrations and failures.

True or False

16. ______ Most pure metals and alloys are cheaper to use than steel.

17. ______ The best use of plastics for corrosion control purposes has been in protective coatings.

18. ______ A common corrosion control method is to use a reactive metal as a coating for a metal with desirable structural properties.

19. ______ Coatings are intended to isolate metal surfaces from electrolytes.

20. ______ The temperature of a surface to be coated should always be below the dew point in order to avoid condensation.

21. ______ When inhibitors are used, corroding substances are neutralized.

22. ______ Permeability can be reduced when formations are treated with inhibitors.

23. ______ Weighted inhibitors are designed to be released upon contact with bottomhole water.

24. ______ Layering oil over brine in storage tanks has proven to be an effective way to prevent oxygen entry.

25. ______ Scavenging, vacuum deaeration, and countercurrent stripping are desirable methods of oxygen removal because of their low costs.

5

Cathodic Protection

OBJECTIVES

Upon completion of this section, the student will be able to:

1. Define *cathodic protection.*
2. List three properties required of anodes.
3. Give two commonly used sources of power for cathodic protection systems and explain how each one works.
4. Explain the effect of current density on cathodic protection systems.
5. Give four reasons why anodes are sometimes packed in coke breeze.
6. List three metals used in galvanic anodes and give the beneficial characteristics of each one.
7. Explain the difference between *impressed current* anodes and *galvanic* anodes.
8. Describe the flow characteristics of cathodic protection currents.
9. Describe the effects of cathodic protection on protective coatings.
10. Explain how cathodic protection currents can be measured.

EFFECTIVENESS

A valuable weapon against corrosion is a process known as *cathodic protection.* This process, by creating and controlling the action of large corrosion cells, has proven particularly effective in preventing corrosion failure on pipelines and offshore structures, resulting in tremendous cost savings for the petroleum industry. Because of its complexity, cathodic protection deserves to be considered apart from the other methods of corrosion control.

THE CATHODIC PROTECTION PROCESS

Corrosion cells act on metal in two ways: cathodic areas of a metal surface in contact with an electrolyte do not corrode; those that become anodic do corrode. Since strong cells tend to overcome weak cells that would otherwise be active in localized areas on large cathodic areas, corrosion can be controlled by strengthening the beneficial cell action and neutralizing the detrimental action. Cathodic protection, in effect, makes the entire surface of an affected structure cathodic, gathers up all the anodes, and transfers them to surfaces where their damage can be tolerated. The anodes are kept corralled by an electromotive potential greater than their own, so the cathode becomes even more cathodic. A structure is said to be *cathodically protected* when all of its surfaces contacting an electrolyte collect current from the electrolyte so that the entire exposed surface becomes a single cathodic area.

Such protection is limited to surfaces that are exposed to the same electrolyte in which the anodes are immersed. The insides of vessels cannot be protected by applying cathodic protection to the outside surfaces. In fact, cathodic protection on the outside of well casings can conceivably cause corrosion on the inside, if the current flows through the annulus fill to the tubing.

SOURCES OF POWER

Although corrosion cells inherently provide cathodic protection to some metal surfaces, additional power is needed in cathodic protection systems to *backout,* or overcome, the positive electrical potentials of anodic areas. Power is also required to force the substitute anodes to deliver current to the electrolyte. A great amount of study and experimentation was needed to develop anode materials that would (1) provide the necessary areas of contact with electrolytes, (2) last long enough to be economical, and (3) minimize power requirements.

Cathodic protection is brought about by connecting the negative terminal of a direct current source to the structure to be protected, or made cathodic. The positive terminal is connected to an electrically conductive substance to be sacrificed, or made anodic. Both must be immersed in the same electrolyte.

Two sources of power are commonly used. The first is DC power. AC power, usually available from power lines, can be reduced by transformers to suitable ranges and converted to DC power by *rectifiers* or motor generators. DC power may also be supplied by solar cells or wind-driven generators when other sources are not available. The obvious disadvantage of these secondary sources is that they are unable to supply power continuously.

The other source utilizes the power stored in metals like magnesium, zinc, and aluminum during their reduction from ores to metals. As shown previously in the electromotive force tables, these metals have galvanic potentials that make them anodic when coupled to steel in an electrolyte. The condition of the steel surfaces, the purity of the metals, and the composition of the electrolytes affect the voltage; still, values of 0.5 V for zinc, 1.0 V for magnesium, and 0.8 V for aluminum are common. Each metal has an electrochemical equivalent, which is the amount of metal carried into solution by discharge of a unit of current from metal to electrolyte. Some approximate chemical equivalents, expressed in lbs/amp per year, are 20 lbs for iron, 17 lbs for magnesium, 26 lbs for zinc, and 6.8 lbs for a special commercially available aluminum alloy. The amount of power available from such anodes is small but often adequate. Because these anodes are used up, they are often called *sacrificial anodes.* (Other anodes are used up also, but usually at much lower rates.)

The amount of DC power used by single cathodic protection units varies widely. In some cases 10 V might be required to drive 1 amp, while in others, 1–3 V might drive 100 amps. Rectifiers have adjustable voltage ranges, and many are supplied with capacities of 10–20 V and 10–100 amps. Overall, efficiencies of rectifiers range from 50 to 70 percent. Galvanic anodes with driving forces in the 0.5–1.0 V range have current outputs per unit that are often measured in milliamps. Efficiencies of galvanic anodes are reduced by local anode action that does not produce usable current. Efficiencies in the 50–80 percent range are common.

POWER REQUIREMENTS

The effective part of the power equation $I = E/R$ is the current, or amperage (I). Knowing the current requirements is the starting point for design of a cathodic protection system.

Every installation is different, but published data are sometimes available to guide systems designers. For bare steel in seawater, for instance, a rate of 4 milliamps per square foot of surface can be used to estimate current requirements. Bare, buried pipelines can be assumed to require 1 milliamp per square foot of surface. At that rate, 1 mile of 30-inch diameter pipe would require about 50 amps. The amperage can be raised by increasing the DC voltage (E), and power input to the cathodic protection unit can be reduced by lowering the circuit resistance (R). In some cases, temporary anodes and power sources are used for preliminary tests to determine power requirements more precisely.

Control of the power source affects protection. DC current flow must somehow be controlled to avoid input power waste and to maintain adequate protection. AC power supplies to rectifiers can be changed, using one of several taps to different parts of secondary stepdown transformer coils. Galvanic anode output can be controlled with electrical resistors in the circuits.

Power requirements may also vary with weather conditions in the case of structures that contact soils. As soil dries, circuit resistance and current requirements increase, because oxygen has greater access to protected surfaces. To remedy this condition, automatic, constant-current rectifiers may be used to raise or lower voltages to correspond with changes in DC circuit resistance.

IMPRESSED CURRENT ANODES

Unlike galvanic, or sacrificial, anodes that transmit power directly to electrolytes, generated power must be introduced into electrolytes. Anode beds, or *ground beds,* are used to protect pipelines and other susceptible structures. Ground beds originally consisted of masses of junk pipe, rails, large castings, or other cheap iron articles buried in the ground, and although they served their purpose, they became outmoded because of their poor efficiencies, their short lives, and their dependence on heavy equipment for installation.

The shape of impressed current anodes has a significant effect on their performance. Anodes that are long and slender, for example, can corrode away near their electrical connections, rendering large portions useless. Cylindrical shapes seem to be the best. The ideal diameter-to-length ratio varies with different materials and conditions. The availability of space and the shape of the structures to be protected may also influence the choice of anode shape. One disadvantage of the cylindrical shape is the *end effect,* by which more current tends to be discharged from the cylinder ends because the greater volume of electrolyte near these surfaces causes more rapid consumption of the anode ends. This phenomenon creates problems with conductor wire connections, since they are usually placed at the ends.

Anodes should have low electrical equivalents, uniform composition, enough mass to give from ten to twenty years of service, a shape that assures complete consumption, and surface areas that reduce circuit resistance and thus use power efficiently. This circuit resistance is more pronounced in the vicinity of anodes, and current density should be kept low to minimize this resistance. Commercially available anodes of compressed and baked graphite, high-silicon cast iron, or magnetite usually meet such requirements (fig. 26). They are usually cylindrical in shape with an insulated wire connection at one end.

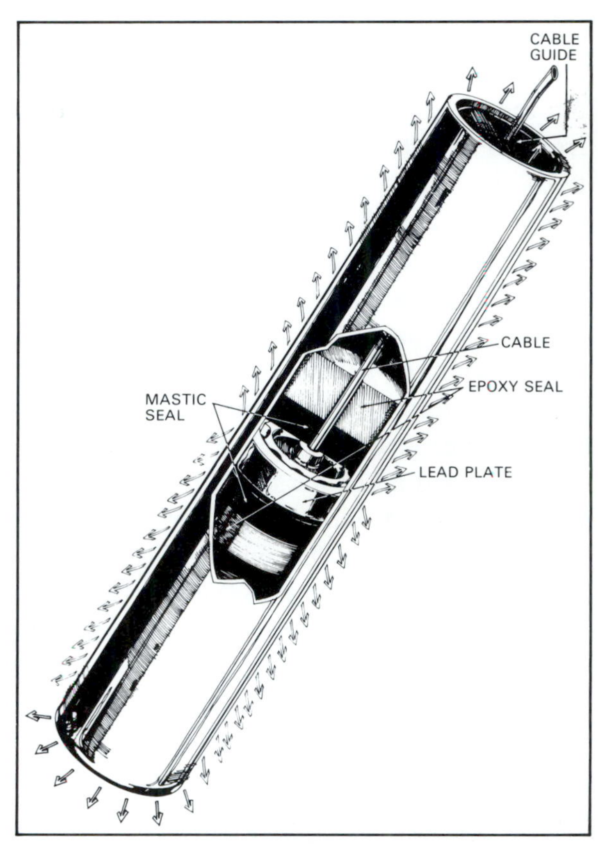

Figure 26. Impressed-current anode with end connection (Courtesy Durco)

When impressed current anodes are installed underground, they are surrounded by packed *coke breeze,* or crushed coke. The coke breeze is desirable for several reasons: it is used up at a low rate because of its low electrochemical equivalent; it provides a uniform environment for the anode; it increases the effective anode size; and it reduces overall costs because it is a comparatively low-cost material.

Obviously, the choice of metal used in anodes is significant. From its position in the electromotive series, platinum is the metal least affected by current discharge in an electrolyte. Accordingly, it is used as a thin covering for titanium. Platinum is also used in combination with columbium. Platinum rods ($^3/_8$ inch × 6 inches) are commonly used inside heater-treaters, separators, filters, and saltwater disposal tanks. They can be installed through the wall with insulated lead wires. Lubricator-type fittings permit installation and changing of the anodes while the vessels are in service. Anode action can be stifled by some inhibitors and emulsion breakers, so test installations should be carried out.

Platinum-titanium anodes of the type shown in figure 27 can protect ship hulls and other areas where bulky anodes would interfere with operations and be difficult to maintain. The shields of such anodes may be formed to fit various shapes, and their hollow construction provides a large area-to-weight ratio, low anode-to-electrolyte resistance, and inside center connections minimizing the end effect. However, the high cost of such materials limits their use to special circumstances.

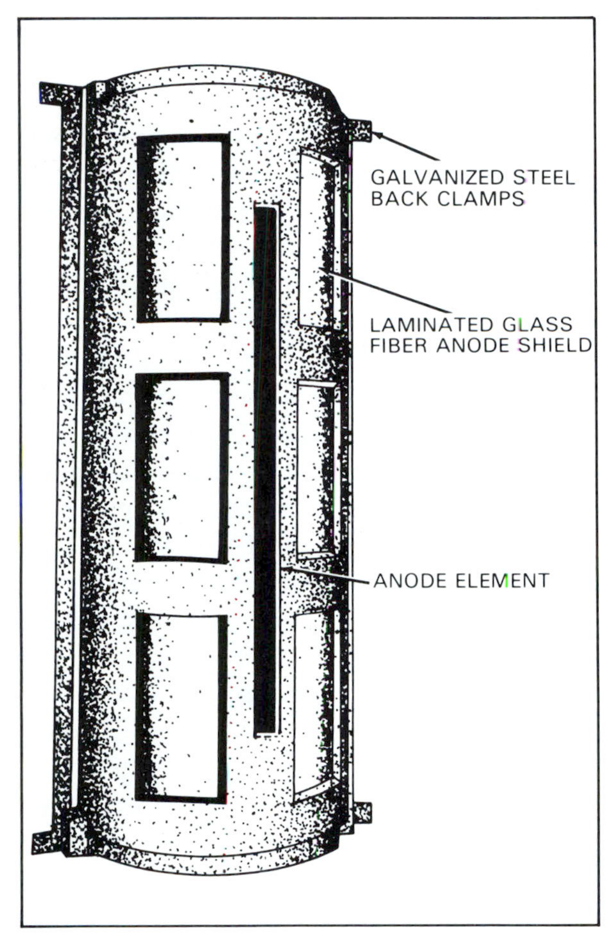

Figure 27. Platinum-titanium anode

Typical cathodic protection units, consisting of a rectifier and a ground bed buried in soil, require several anodes to provide low circuit resistance and reasonably long ground bed life. Frequently the individual anodes are spaced about 15 feet apart in a straight line (fig. 28). In more conductive electrolytes such as seawater or salt-marsh soil, wide spacing is not so essential, but it

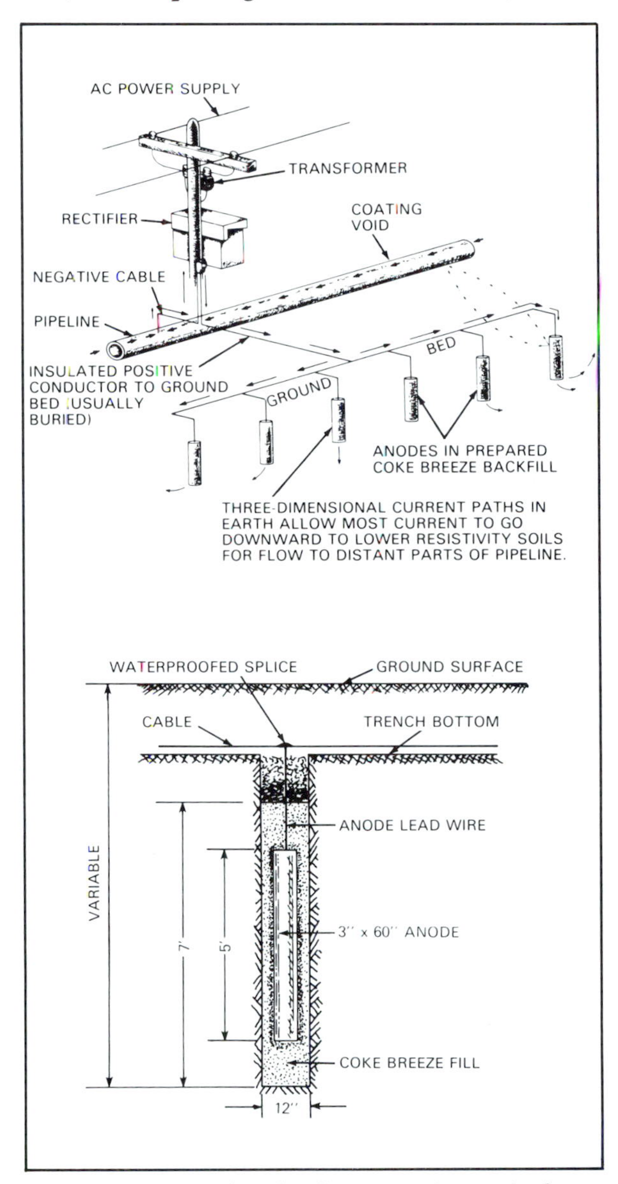

Figure 28. Typical cathodic protection unit showing anode installation

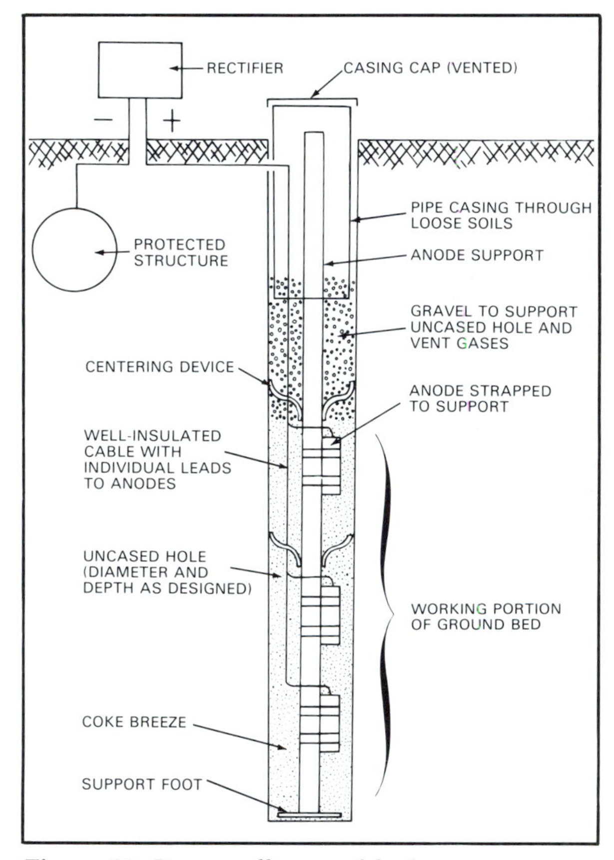

Figure 29. Deep well ground bed

may be desirable in order to reduce an inrush of current to nearby portions of the protected structure.

Moisture penetration, which causes rapid corrosion and high electrical resistance, makes electrical connections to anodes difficult. Portions of graphite anodes around connections are impregnated with paraffin or other substances to resist moisture. High-silicon cast iron anodes in tubular form have the electrical connection inside at the midpoint, protecting the connection and improving the surface-to-weight ratio. These anodes should be handled carefully to avoid damage to connections and insulation on positive conductors.

Deep well ground beds are often used to reach lower resistivity strata or when surface areas for construction are limited (fig. 29). Such beds may also provide better distribution of current to protected structures. Carbon dioxide gas is a product of the electrochemical action at graphite or coke anodes. For vertically installed anodes, gravel should be used to backfill the holes above the carbonaceous material to allow gases to escape. This venting is required in deep ground beds to avoid the entrapment of gases that could exclude moisture from anodes and raise their electrical resistance. Casing caps should also be vented.

It may be desirable to suspend anodes in electrolytes such as seawater or fresh water in metal tanks. In such cases, bare rods of graphite or silicon cast iron may be used. Dissolved iron can be a contaminant in fresh water, so iron anodes should not be used.

GALVANIC ANODES

Galvanic anodes have low current outputs per unit, but there are many situations for which they are suitable, and they are used extensively because of certain advantages they offer. They can be placed quite close to the surface they are designed to protect. Their small voltages and their direct connection to protected structures practically eliminate the risk of sparking at broken connections in combustible atmospheres. In some cases, the performance characteristics of these anodes are so well known that they can be attached to structures that are under construction. Anodes are routinely welded to offshore platforms. The weight and size of the anodes and their placement on the structure can be specified to provide protection for a definite period.

Well-insulated structures such as coated pipelines or lined tanks require small amounts of protective current, which can often be supplied by comparatively maintenance-free galvanic anodes. The anodes may be distributed along pipelines near other structures so that there is little risk of cathodic interference by stray currents, or they may be used in crowded areas to obviate the shielding effect of structures between anodes and protected structures. Galvanic anodes require little auxiliary equipment, so they can be used where structures such as rectifiers cannot.

Magnesium

Magnesium is the metal most often used for galvanic anodes because of its high driving potential of 1 V. It can be used underground, in seawater, and in oil field brines. Magnesium is ordinarily cast around a steel rod that protrudes at one end, providing a connection to the structure to be protected. The rod, an electrical conductor, also provides continued service if unequal current discharge uses up all the magnesium in localized areas, separating the unused portions.

Galvanic anodes buried in soil are especially subject to this uneven use, and special backfills containing gypsum can be used to provide more uniform surroundings. The gypsum may be mixed with clay, but it should be intimately mixed for uniformity. The gypsum, which dissolves slowly, provides a low-resistivity environment for the anode and attracts moisture. The clay helps hold the moisture in.

The much-used 17-lb anodes are often prepackaged with backfill. Anodes up to 5 feet long and 3 or more inches in diameter are usually installed in holes that are bored in the ground and backfilled with the gypsum-clay mixture (fig. 30). Since magnesium is consumed at the rate of 17 lbs/amp per year, the output of a 17-lb anode would need to be limited to 0.1 amp for a ten-year life. In many cases, clusters of larger-sized anodes are needed to obtain the required current and the desired service life.

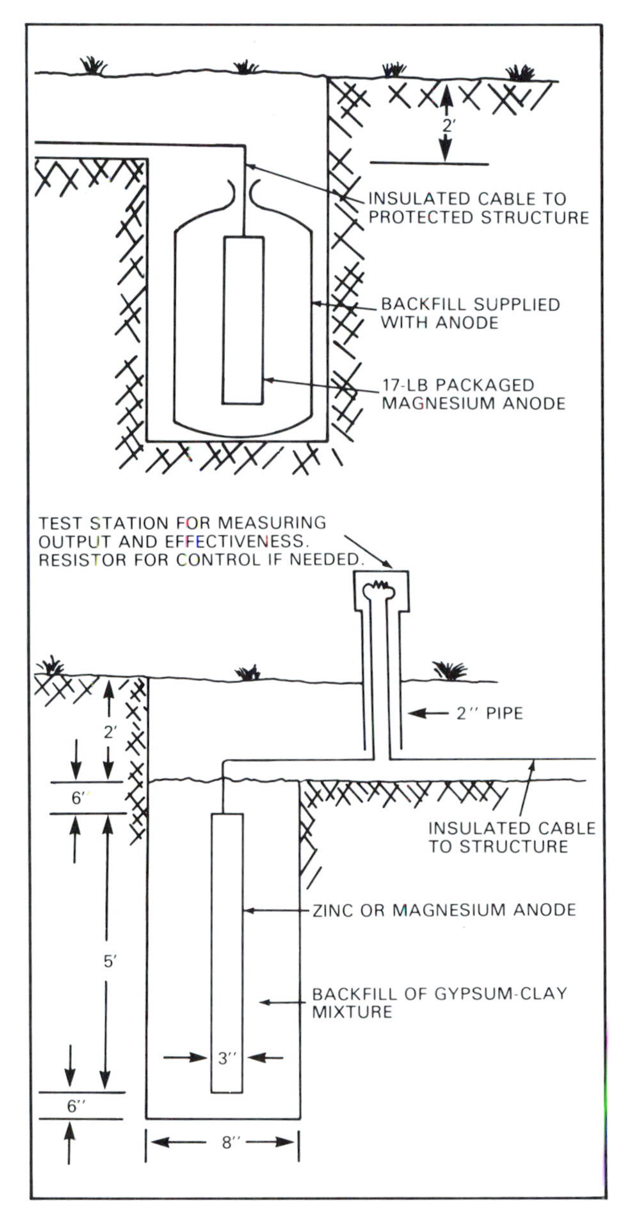

Figure 30. Examples of galvanic anode installations

Zinc

Zinc has long been used for cathodic protection. It is most useful in low-resistivity soils and waters. In high-resistivity soils, the low driving voltage reduces effectiveness. When carbonates are present, scales can

Figure 31. A 725-lb anode attached to offshore platform structure

build up, isolating the carbonates and, in some cases, reversing their polarity with respect to steel. Usually it is advisable to use prepared backfill when zinc is used in soil.

Hemicylindrical zinc bracelets are often welded at intervals along subsea pipelines that are sheathed in concrete for extra weight. These anodes do not interfere with construction, and they may be designed to provide protection for definite periods such as twenty or thirty years.

Aluminum

Aluminum is not very effective in soils, but special aluminum alloys have shown good resistance to the formation of insulating films in seawater and brines. Single anodes weighing more than 700 lbs have been attached to offshore platforms (fig. 31). Anodes may be attached by divers, also (fig. 32). With their low electrochemical equivalents, aluminum anodes can be designed for long service life.

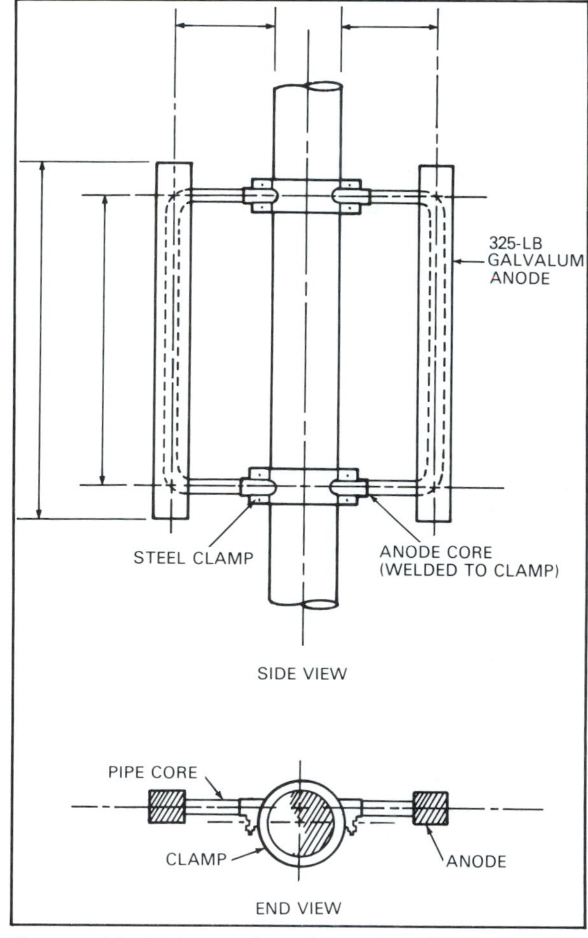

Figure 32. Schematic of anode placement on underwater structure

FLOW CHARACTERISTICS OF CATHODIC PROTECTION CURRENTS

The characteristics of current flow through electrolytes from anode to cathode govern the positioning of anodes with respect to protected surfaces. Like all natural forces, protective currents follow lines of least resistance, and current travels across the most direct path between anode and cathode.

However, since cathodic surfaces become polarized and resist entry of more current, protected areas tend to spread until equilibrium is reached. Spreading continues until the $E=IR$ relationship stablizes. (E is the applied voltage between anode and cathode; I is the current; and R is the total resistance of the circuit.) When the potential of the electrolyte to the protected structure is too great, hydrogen gas bubbles will evolve at the cathode. This reaction wastes energy and usually damages protective coatings.

Current tends to flow in three-dimensional paths away from anodes. This flow pattern can be affected by the nearness of anodes to electrolyte surfaces, nonuniform conductivity of electrolytes or current paths, and crowding from the current output of other anodes. Compared to metals, electrolytes have high electrical resistances, and the IR drop in an electrolyte in the vicinity of anodes is much higher than at greater distances from the anodes. In massive electrolytes offshore, the IR drop soon reaches 0, so the resistance of electrolytic paths might be the same for objects 500 feet or 5 miles apart. The resistances of the metallic paths would, of course, be different. If parts of the cathode are close enough to anodes to be in zones of high current density, the current may crowd into this low-resistance path, causing overprotection and wasting power. Anodes should be separated from protected surfaces according to applied voltages and current outputs.

The condition of cathodic surfaces also affects current flow. Surfaces with good high-resistance coatings may have such a low inflow of current per unit of area that anodes can be spaced fairly close together. Close spacing is usually unavoidable when galvanic or impressed current anodes are installed in enclosed vessels like heat exchangers or heater-treaters where projections into electrolytes are not practical. In such cases, anodes may be mounted quite close to the surfaces to be protected, sometimes in the form of plates contoured to fit those surfaces. Sheets of micarta, rubber, or some other type of insulation should be used to shield and provide extra space around the protected surface under the anode. Whenever practical, anodes should be centered within electrolytes or arranged for uniform distribution of current.

Anodes, either galvanic or impressed current, may be suspended inside vessels such as tanks or in water to protect platforms or other structures. In very conductive waters such as brines, the current output of individual galvanic anodes may need to be restricted. Casting anodes in the form of spheres provides the highest possible mass-to-surface ratio, helps restrict current output, and eliminates the end effect.

Usually, insulated conductors should be installed through vessel shells so that voltage and amperage can be measured. In some cases this may not be necessary, and galvanic anodes may be attached directly to protected structures. Direct attachment should be safe when operating conditions are stable and when the effective life of the anodes is known. Occasionally vessels are opened for inspection.

The tendency of current flow to follow the path of least resistance can reduce or eliminate the possibility of cathodic protection for clusters of objects. For instance, the outside tubes in tube bundles of heat exchangers absorb most of the current intended for their external protection. The inside surfaces of pipes and tubes cannot ordinarily be cathodically protected. Protective current "thrown" at the open ends of tubes in a tube sheet cannot travel far through electrolytes into the tubes because electrolyte resistance soon equals that of the metallic paths. Anodes can be installed inside very large pipes without causing too much interference with the flow of the pipe contents.

Several parallel buried pipelines can be protected without difficulty, provided they are well coated. In the case of bare or poorly-coated pipe, the outside lines collect large proportions of the current, leaving smaller and perhaps inadequate amounts for the inside lines. Pipe coatings are so widely used that large quantities of such bare pipe are not likely to be encountered, but supplying more current or installing anodes at considerable depth (preferably under the pipelines) can remedy the problem when it exists. Of course, pipelines can be replaced, repaired, or recoated, but only at great expense.

Intricate networks of buried pipe such as those at treating plants or refineries pose certain problems with current distribution. In new construction, the use of overhead piping can be a way of avoiding the problem. Protective coatings that are resistant to solution with solvents that might be spilled can be used in conjunction with galvanic anodes. Deep ground beds usually provide adequate current distribution, also.

When several oil or gas wells in a field are to be cathodically protected, designers are often tempted to install a few large cathodic protection units, each serving several wells. This arrangement offers convenient, economical construction, plus easier maintenance and operation. Negative conductors are run to each well, and the amount of current collected is adjustable, but maintaining the proper balance of current collected from each well is difficult. Because this balance is rarely achieved in a satisfactory manner, it seems best to provide each well with its own unit.

Nonconductive materials can also interfere with protective current flow. Massive, high-resistance rocks can prevent full protection of sections of pipelines. When layers of protective coatings become detached, water can infiltrate, blocking the flow of protective current and allowing corrosion to proceed.

PROTECTIVE COATINGS WITH CATHODIC PROTECTION

Protective coatings, even poor ones, have a marked effect on current requirements, as shown in table 4. Service conditions such as high temperatures make it impractical to coat some surfaces, but most surfaces that need cathodic protection also need coatings.

In choosing a coating, the effect of cathodic protection should always be considered. Water-permeable coatings may suffer cathodic disbonding because of the hydrogen films deposited on metal surfaces. If cathodic voltage is high enough to cause molecular hydrogen bubbles to evolve, coatings are usually lifted off quickly. Because cathodically protected surfaces also

TABLE 4
Current Required for Protecting
10 Miles of 36-Inch Diameter Pipe
under Assumed Conditions

Effective Coating Resistance For One Square Foot (ohms)	Current Required (amps)
Bare Pipe*	500
10,000	14.91
25,000	5.964
50,000	2.982
100,000	1.491
500,000	0.2982
1,000,000	0.1491
5,000,000	0.0298
Perfect Coating	0.000058

NOTE: Perfect coating is assumed to be holiday-free and to consist of a $^3/_{32}$-inch layer of material having a resistivity of 10^{13} ohm-centimetres. Current required is that needed to cause a 0.3-V drop across the effective resistance between the pipeline and remote earth. Polarization effects are neglected. Lower values of coating resistance result from quantities and degrees of imperfections in the coatings.
*Bare pipe is assumed to require a minimum of 1 milliamp of current per square foot.

tend to become alkaline, coatings need to be alkali-resistant. Saponifiable oils can be converted into soap.

In some environments such as seawater, brines, and certain soils, cathodic protection causes a hard, calcareous scale to form. This scale, which is a fairly good coating, results from the decomposition of water, according to the formula

$$2H^{+} + 2(OH)^{-} - e \longrightarrow 2(OH)^{-} + H_2 .$$

The surplus hydroxyl ions react with calcium and magnesium bicarbonates to form carbonates, which are powdered solids.The deposit of silicates tends to bind the products of this reaction into hard, electrically resistant scales. These white scales can be seen on surfaces that have been protected for considerable periods of time.

DETERMINING CATHODIC PROTECTION

The best way to determine whether a metal surface is cathodically protected is to measure the difference in electrical potential between that surface and the surrounding electrolyte. A protected surface will be negative to the electrolyte, indicating a flow of current from electrolyte to metal.

As mentioned earlier, copper sulfate half-cells are most often used for measuring these potential differences. The most commonly accepted negative value for complete protection of steel is −0.85 V, as measured by copper sulfate half-cell. This measurement consists of the 0.55-V natural difference between noncorroding steel and the copper sulfate half-cell, plus 0.3 V caused by inflow of current. Experience has shown that naturally occurring corrosion cells have voltages of less than 0.3 V and that they may be overcome by this voltage. To indicate complete protection, the criterion of a negative potential shift of 0.3 V to −0.85 V is sometimes used. There is some disagreement about this second criterion, but it is adequate in the vast majority of cases.

Placement of the copper sulfate half-cell is important; the closer the half-cell is placed to the metal surface without touching, the better. Current flowing through a comparatively high-resistance electrolyte results in *IR* drop, or voltage change. Part of the voltage difference between the steel surface and the half-cell can be caused by this current flow. For structures buried in soil, the half-cell should be positioned directly over the structure (fig. 33).

Current flow patterns in soils, resulting from corrosion or from cathodic protection, can be mapped by using two half-cells. To alleviate the effects of current flow on potential measurements, the half-cells can

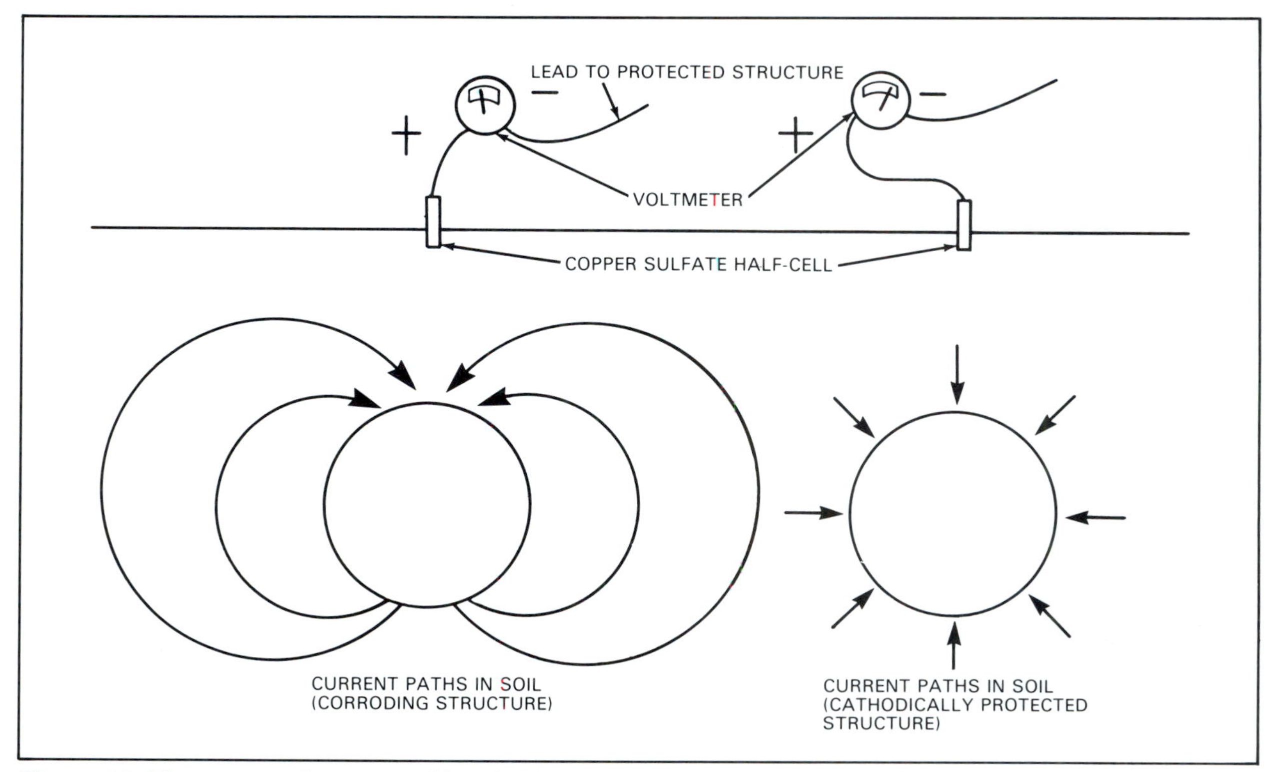

Figure 33. Placement of copper sulfate half-cell to minimize effect of soil currents

be placed at considerable distances from the structure. When no change in potential can be detected by moving the half-cell farther away from the protected structure, half-cell contact points are considered to be sufficiently remote.

In measuring natural potentials (those not influenced by intentionally introduced current) it should be remembered that the half-cell is influenced by the condition of the metal surface it "sees" (when the half-cell is thought of as a camera). Since cathodes are large and anodes are small, what the half-cell sees is likely to be mostly cathodic. In the case of buried pipe, for instance, corrosion tends to be greatest on the bottom of the pipe. The associated currents flow mostly in circumferential directions to the top of the pipe. Half-cells located above the pipe will see only cathodes. Surveys made before cathodic protection begins often show the areas of greatest corrosion as being the most negative. Any shift to more negative values resulting from cathodic protection currents is beneficial, but full protection is not assured unless the criterion of -0.85 V to a copper sulfate half-cell can be achieved.

The direction and magnitude of current flow in pipelines can also be measured by observing the polarity and *IR* drop between points on the pipeline. Current magnitude can be calculated by measuring voltage change between two points on a pipe with the formula $I = E/R$. (In this case, R is a function of the resistivity of the steel involved and the weight of the pipe per unit of length.) Figure 34 demonstrates how such measurements can be taken. First, the voltage between contact points A and B is measured. Then the voltmeter and contacts are moved along the pipe to locate anodic areas. The direction of current flow in the

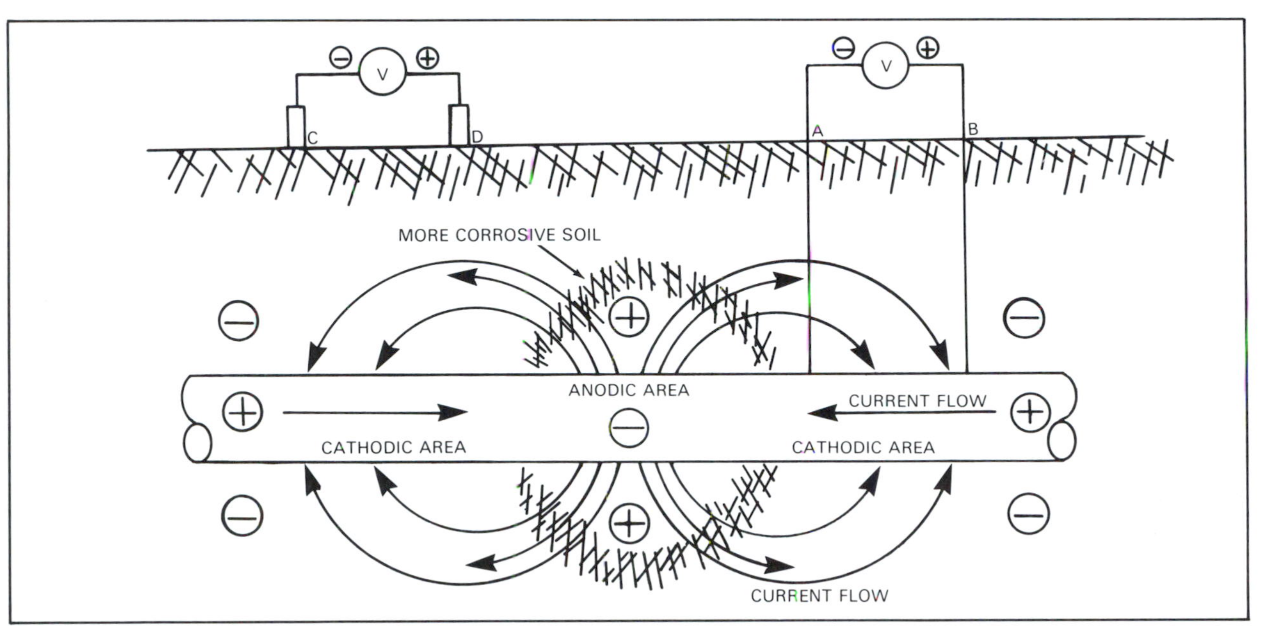

Figure 34. Measurement of current flow in a pipeline using two copper sulfate half-cells

soil and the approximate amount of current can be determined by using the voltmeter and two copper sulfate half-cells at points *C* and *D*. Mapping of current flow patterns in pipe and soil helps to locate corroding areas or hot spots and indicate suitable locations for anode placement.

Some contact methods can also be used with suitable equipment to find areas of current discharge and collection in downhole piping. For galvanic anodes inside vessels with insulated, through-the-wall connections, current output is usually an adequate measurement of performance. Reductions in output indicate that anodes need to be cleaned or replaced. Vessels such as heat exchangers and condensers have parts that need internal cathodic protection: steel shells, cast iron or cast steel heads, fire tubes, and tube sheets. These internal spaces are not readily accessible for metal-to-electrolyte potential measurements. Performance records and voltage and amperage measurements can help measure the effects of anode use.

CATHODIC PROTECTION INTERFERENCE

Currents introduced into large electrolytes by cathodic protection may act as stray currents to structures that are not meant to be affected (fig. 35). To minimize the likelihood of such interference, care should be exercised in the placement of anodes, especially the impressed current

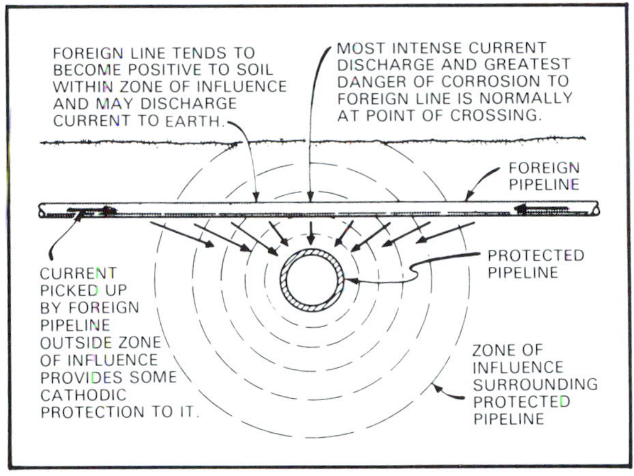

Figure 35. The effect of potential gradients surrounding a cathodically protected bare line on a foreign pipeline

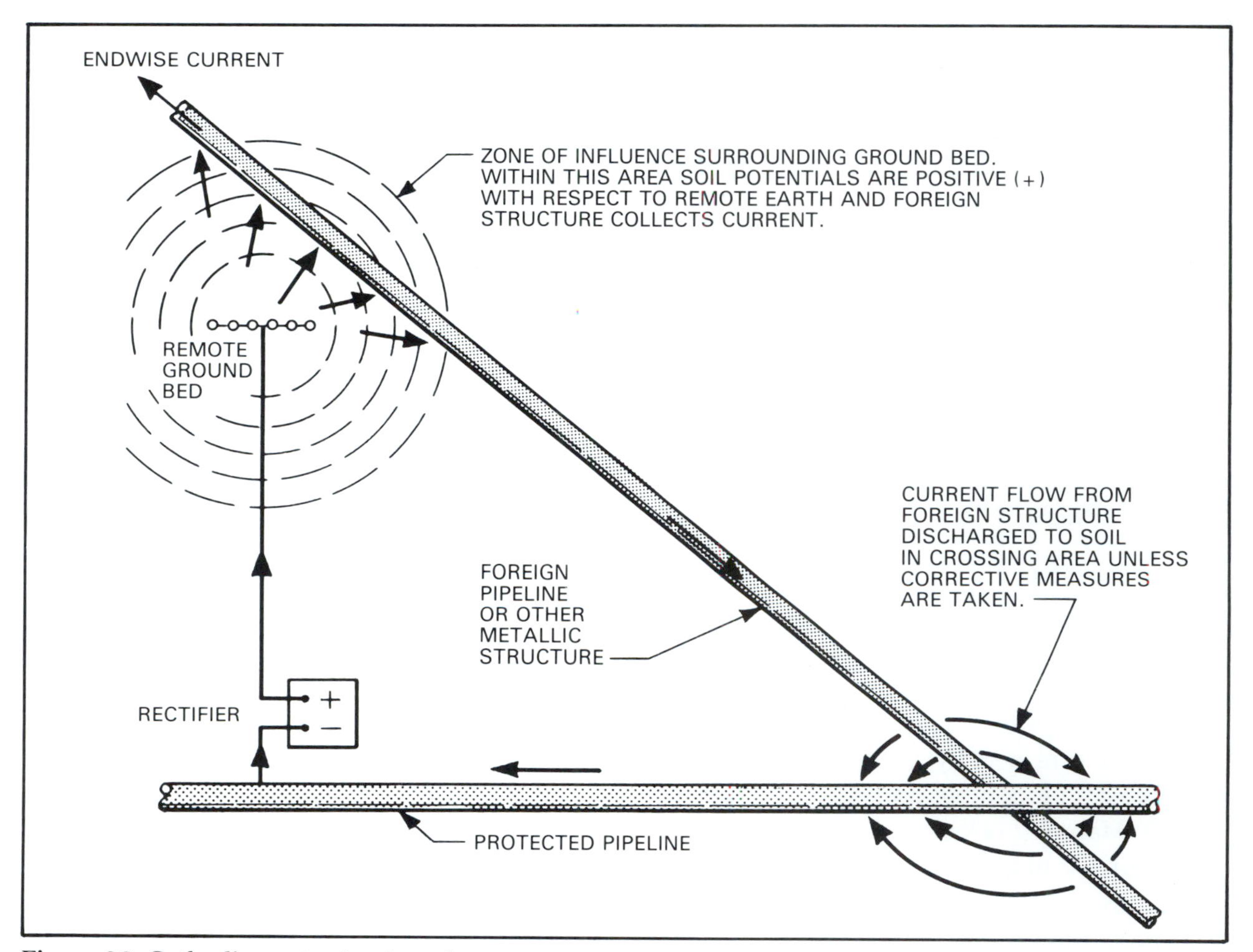

Figure 36. Cathodic protection interference accentuated by undesirable ground bed location.

type (fig. 36). Interference is sometimes inevitable, but cooperation between owners of interfering structures can help. Joint tests, before and after, can be made to compare normal structure-to-electrolyte potentials with the potentials registered after cathodic protection begins. The usual remedy is to install bonds between structures to restore foreign structures to their preprotection potentials. The electrical resistance of these bonds can be varied to adjust potentials. Foreign structures may require considerable amounts of protective current, a possibility that should be taken into account when designing protection systems.

Problems arise when a new, well-coated structure must be installed near an existing structure that is bare or poorly coated. The older structure may require much more current through the bond than the newer one requires for complete protection. In such cases, the well-coated pipeline will require so little current that the amount at any particular point will be extremely small, and the effect on nearby structures will be minimal. The ground bed for the new structure should be as remote from the old structure as is convenient in order to eliminate the need for a bond or at least minimize the exchange of current. Pipelines are good examples because of their simple structures.

Other structures would have similar but fewer interference problems.

THE FUTURE OF CORROSION CONTROL

Oil and gas production is always facing new problems, creating new challenges in corrosion control. With wells now being drilled to depths of more than 30,000 feet, corrosion control substances must be able to withstand extreme heat and pressure. Also, the use of agents such as water, steam, fire, and polymers in the recovery of residual oil have increased the complexity of control procedures.

Corrosion failure still costs the petroleum industry billions of dollars each year. Corrosion research is always leading to the development of improved control methods, but their effectiveness will still depend on the diligence and cooperation of production personnel.

Self-Test

5: Cathodic Protection

(Multiply each correct answer by three to arrive at your percentage of competency.)

Fill in the Blank

1. A structure is said to be ____________ ____________ when all surfaces that contact an electrolyte collect current from the electrolyte so that the entire exposed surface becomes a single cathodic area.
2. Anodes used in cathodic protection must be able to (1) ____________; (2) ____________; and (3) ____________.
3. ____________ anodes are used up at a faster rate than other types of anodes.
4. The ______ ____________ is one disadvantage of the cylindrical-shaped anode.
5. Current density in an electrolyte should be kept low in order to ____________ ____________.
6. Anodes are sometimes packed in coke breeze, which is desirable because (1) ____________ ____________;
   (2) ____________;
   (3) ____________; and
   (4) ____________.
7. ____________ is the anode material least affected by current discharge in an electrolyte.
8. Graphite anode connections are sometimes impregnated with paraffin in order to avoid ____________ ____________.
9. ____________ ____________ ground beds are often used to reach lower-resistivity strata.
10. Galvanic anodes buried in soil are sometimes backfilled with a gypsum-clay mixture to prevent ____________ ____________.

11. ____________________ anodes have shown good resistance to insulating films in seawater and brines.

12. Even though cathodic currents follow the most direct path from anode to cathode, protected areas tend to spread because ____
____.

13. The *IR* drop in an electrolyte in the vicinity of anodes is much higher than at greater distances, because ____
____.

14. The best way to determine if a metal surface is cathodically protected is to measure the difference in electrical potential between ____
and ____.

15. The magnitude of cathodic protection currents can be calculated by measuring the voltage change between two points on a pipe using the formula ___________.

True or False

16. ______ Cathodic protection eliminates the corrosion process.

17. ______ The insides of vessels can often be protected by applying cathodic protection to the vessel's outside surfaces.

18. ______ The two most commonly used sources of power for cathodic protection are AC power and DC power.

19. ______ Power requirements for cathodic protection systems are relatively constant.

20. ______ Rectangular anodes seem to be the most efficient type.

21. ______ One advantage of galvanic anodes is that they can be placed quite close to the surfaces to be protected.

22. ______ The tendency of current to follow the path of least resistance can reduce the possibility of cathodic protection for clusters of objects.

23. ______ When several wells are to be protected, a few large cathodic protection units can provide the most efficient protection.

24. ______ Most protective coatings work well with any cathodic protection system.

25. ______ Stray currents can be caused by cathodic protection units.

Glossary

A

acid *n:* any chemical compound, one element of which is hydrogen, that dissociates in solution to produce free hydrogen ions. For example, hydrochloric acid, HCl, dissociates in water to produce hydrogen ions, H^+, and chloride ions, Cl^-. This reaction is expressed chemically as $HCl \rightarrow H^- + Cl^-$. See *ion.*

acid brittleness *n:* low ductility of a metal due to its absorption of hydrogen gas. Also called hydrogen embrittlement.

acidize *v:* to treat oil-bearing limestone or other formations with acid for the purpose of increasing production. Hydrochloric or other acid is injected into the formation under pressure. The acid etches the rock, enlarging the pore spaces and passages through which the reservoir fluids flow. The acid is held under pressure for a period of time and then pumped out, after which the well is swabbed and put back into production. Chemical inhibitors combined with the acid prevent corrosion of the pipe.

aerobic bacteria *n pl:* bacteria that require free oxygen for their life processes. Aerobic bacteria can produce slime or scum that accumulates on metal surfaces, causing oxygen-concentration cell corrosion.

alkali *n:* a substance having marked basic (alkaline) properties, such as a hydroxide of an alkali metal. See *base.*

alloy *n:* a substance with metallic properties composed of two or more elements in solid solution. See *ferrous alloy* and *nonferrous alloy.*

alternating current *n:* an electric current that reverses its direction of flow at regular intervals.

anaerobic bacteria *n pl:* bacteria that do not require free oxygen to live or are not destroyed by its absence. Under certain conditions, anaerobic bacteria can cause scale to form in water-handling facilities in oil fields or hydrogen sulfide to be produced from sulfates.

anchor pattern *n:* the pattern of minute projections from a metal surface produced by sandblasting, shot blasting, or chemical etching to enhance the adhesiveness of surface coatings.

anion *n:* a negatively charged ion.

anode *n:* 1. one of two electrodes in an electrolytic cell; represented as the negative terminal of the cell, it is the area from which electrons flow. In a primary cell it is the electrode that is wasted or eaten away. 2. in cathodic protection systems, an electrode to which a positive potential of electricity is applied, or a sacrificial anode, which provides protection to a structure by forming one electrode of an electric cell.

annular space *n:* 1. the space surrounding a cylindrical object within a cylinder. 2. the space around a pipe in a wellbore, the outer wall of which may be the wall of either the borehole or the casing; sometimes termed the annulus.

annulus *n:* also called annular space. See *annular space.*

artificial lift *n:* any method used to raise oil to the surface through a well after reservoir pressure has declined to the point at which the well no longer produces by means of natural energy. Sucker rod pumps, gas lift, hydraulic pumps, and submersible electric pumps are the most common forms of artificial lift.

asbestos felt *n:* a wrapping material, consisting of asbestos saturated with asphalt, which is one element of pipeline coatings.

atom *n:* the smallest quantity of an element capable of either entering into a chemical combination or existing alone.

atomize *v:* to spray a liquid through a restricted opening, causing it to break into tiny droplets and mix thoroughly with the surrounding air.

B

backout *v:* to overcome the positive electrical potentials of anodic areas in cathodic protection systems.

barrel *n:* a measure of volume for petroleum products in the United States. One barrel is the equivalent of 42 U.S. gallons or 0.15899 cubic metres. One cubic metre equals 6.2897 barrels.

base *n:* a substance capable of reacting with an acid to form a salt. A typical base is sodium hydroxide (caustic), with the chemical formula NaOH. For example, sodium hydroxide combines with hydrochloric acid to form sodium chloride (a salt) and water; this reaction is written chemically as

$$NaOH + HCl \rightarrow NaCl + H_2O.$$

base metal *n:* 1. any of the reactive metals at the lower end of the electrochemical series. 2. metal to which cladding or plating is applied.

bitumen *n:* substance of dark to black color consisting almost entirely of carbon and hydrogen with very little oxygen, nitrogen, or sulfur. Bitumens occur naturally, and they can also be obtained by chemical decomposition.

bond *n:* the adhering or joining together of two materials (as cement to formation). *v:* to adhere or to join to another material.

brine *n:* water that has a large quantity of salt, especially sodium chloride, dissolved in it; salt water.

C

caliper log *n:* a record whereby the diameter of the wellbore is ascertained, indicating undue enlargement due to caving in, washout, or other causes. The caliper log also reveals corrosion, scaling, or pitting inside tubular goods.

carbonate *n:* 1. a salt of carbonic acid. 2. a compound containing the carbonate (CO_3^{--}) radical.

casing *n:* steel pipe placed in an oil or gas well as drilling progresses to prevent the wall of the hole from caving in during drilling, to prevent seepage of fluids, and to provide a means of extracting petroleum if the well is productive.

cathode *n:* 1. one of two electrodes in an electrolytic cell, represented as the positive terminal of a cell. 2. in cathodic protection systems, the protected structure that is representative of the cathode and is protected by having a conventional current flow from an anode to the structure through the electrolyte.

cathodic protection *n:* a method of protecting a metal structure from corrosion by making its surfaces cathodic and controlling the location of anodic areas so that corrosion damage can be reduced to tolerable levels.

cation *n:* a positively charged ion.

chemisorption *n:* chemical adsorption.

clay *n:* a fine crystalline material of hydrous silicates, resulting primarily from the decomposition of feldspathic rocks.

coating *n:* in corrosion control, any material that forms a continuous film over a metal surface to prevent corrosion damage.

coke *n:* a cellular solid residue produced from the dry distillation of certain carbonaceous materials, containing carbon as its principal constituent.

coke breeze *n:* crushed coke, used for packing underground anodes in cathodic protection systems to obtain increased anode efficiency at a reduced cost. See *coke.*

compound *n:* 1. a mechanism used to transmit power from the engines to the pump, drawworks, and other machinery on a drilling rig. It is composed of clutches, chains and sprockets, belts and pulleys, and a number of shafts, both driven and driving. 2. a substance formed by the chemical union of two or more elements in definite proportions; the smallest particle of a chemical compound is a molecule. *v:* to connect two or more power-producing devices, such as engines, to run driven equipment, such as the drawworks.

continuous treatment *n:* a method of applying corrosion inhibitors to production fluids, in which the concentration of the inhibitor can be maintained at constant levels.

copper sulfate electrode *n:* a commonly used nonpolarizing electrode used in corrosion control to measure the electrical potential of a metal structure to a surrounding electrolyte to determine the potential for corrosion damage or to monitor the effectiveness of existing control measures. See *half-cell.*

corrosion cell *n:* the pattern of flow of electric current between metals in an electrolyte, which causes metal to corrode or deteriorate.

corrosion control *n:* the measures used to prevent or reduce the effects of corrosion. These practices can range from simply painting metal to isolate it from moisture and chemicals and to insulate it from galvanic currents to cathodic protection, in which a galvanic or impressed direct electric current renders a pipeline cathodic, thus causing it to be a negative element in the circuit. The use of chemical inhibitors and closed systems are other examples of corrosion control.

corrosion coupon *n:* a metal strip inserted into a system to monitor corrosion rate and to indicate corrosion inhibitor effectiveness.

countercurrent stripping *n:* the use of natural or inert gas to remove oxygen from production systems.

current *n:* the flow of electric charge or the rate of such flow, measured in amperes.

D

dew point *n:* the temperature and pressure at which a liquid begins to condense out of a gas. For example, if a constant pressure is held on a certain volume of gas but the temperature is reduced, a point is reached at which droplets of liquid condense out of the gas. That point is the dew point of the gas at that pressure. Similarly, if a constant temperature is maintained on a volume of gas but the pressure is increased, the point at which liquid begins to condense out is the dew point at that temperature.

diffusion *n:* 1. the spontaneous movement and intermingling of particles of liquids, gases, or solids. 2. the migration of dissolved substances to areas of least concentration.

direct current *n:* electric current that flows in only one direction.

dispersible inhibitor *n:* an inhibitor substance that can be evenly dispersed in another liquid with moderate agitation.

dispersion *n:* a suspension of extremely fine particles in a liquid (such as colloids in a colloidal solution).

dissociation *n:* the separation of a molecule into two or more fragments (atoms, ions) by interaction with another body or by the absorption of electromagnetic radiation.

downhole *adj, adv:* pertaining to the wellbore.

E

electrical potential *n:* voltage.

electrochemical series *n:* See *electromotive series.*

electrode *n:* a conductor of electric current as it leaves or enters a medium such as an electrolyte, a gas, or a vacuum.

electrodeposition *n:* an electrochemical process by which metal settles out of an electrolyte that contains the metal's ions and is then deposited at the cathode of the cell.

electrolyte *n:* 1. a chemical that, when dissolved in water, dissociates into positive and negative ions, thus increasing its electrical conductivity. See *dissociation.* 2. the electrically conductive solution that must be present for a corrosion cell to exist.

electromotive force *n:* 1. the force that drives electrons and thus produces an electric current. 2. the voltage or electric pressure that causes an electric current to flow along a conductor. The abbreviation for electromotive force is emf.

electromotive series *n:* a list of elements arranged in order of activity (tendency to lose electrons). The following metals are so arranged: magnesium, beryllium, aluminum, zinc, chromium, iron, cadmium, nickel, tin, copper, silver, and gold. If two metals widely separated in the list (e.g., magnesium and iron) are placed in an electrolyte and connected by a metallic conductor, an electromotive force is produced. See *corrosion.*

emf *abbr:* electromotive force.

epoxy *n:* any compound characterized by the presence of a reactive chemical structure that has an oxygen atom joined to each of two carbon atoms that are already bonded.

exchanger *n:* a piping arrangement that permits heat from one fluid to be transferred to another fluid as they travel countercurrent to one another. In the heat exchanger of an emulsion-treating unit, heat from the outgoing clean oil is transferred to the incoming well fluid, cooling the oil and heating the well fluid.

F

fast-release inhibitors *n:* inhibitor substances that are released immediately upon introduction into a system.

ferrous alloy *n:* an alloy that contains iron as the largest single constituent.

fouling *n:* the buildup of deposits of marine organisms or corrosion products on metal surfaces.

frp *abbr:* fiberglass-reinforced plastic.

G

galvanic anode *n:* in cathodic protection, a sacrificial anode that produces current flow through galvanic action.

galvanic cell *n:* electrolytic cell brought about by the difference in electrical potential between two dissimilar metals.

galvanic corrosion *n:* a type of corrosion that occurs when a small electric current flows from one piece of metal equipment to another. It is particularly prevalent when two dissimilar metal objects are joined together in an environment in which electricity can flow such as two dissimilar joints of tubing in an oil or gas well.

ground bed *n:* in cathodic protection, an interconnected group of impressed current anodes that absorbs the damage caused by generated electric current.

Gunite *n:* trade name for a cement-sand mixture used to seal pipe against air, moisture, and corrosion damage.

H

half-cell *n:* a single electrode immersed in an electrolyte for the purpose of measuring metal-to-electrolyte potentials and, therefore, the corrosion tendency of a particular system.

heat-exchanger *n:* See *exchanger.*

heater-treater *n:* a vessel that heats an emulsion and removes water and gas from the oil to raise it to a quality acceptable for pipeline transmission. A heater-treater is a combination of a heater, free-water knockout, and oil and gas separator.

high-purity water *n:* water that has little or no ionic content and is therefore a poor conductor of electricity.

hot spot *n:* an abnormally hot place on a casing coupling when a joint is being made up. It usually indicates worn threads on the pipe and in the coupling.

hydrogen embrittlement *n:* also called acid brittleness. See *acid brittleness.*

hydrogen patch probe *n:* an instrument that, when attached to the exterior of a vessel that has been corroded by hydrogen sulfide, senses the hydrogen content in the steel and records the rate of corrosion.

hydrogen sulfide *n:* a gaseous compound of sulfur and hydrogen, H_2S, commonly found in petroleum, which causes the foul smell of sour petroleum fractions. It is extremely poisonous and corrosive.

I

impressed current anode *n:* anode to which an external source of positive electricity (as from a rectifier, DC generator, etc.) is applied. The negative electricity is applied to the pipeline, casing, or other structure to be protected by the impressed current method of cathodic protection.

inhibitor *n:* an additive used to retard undesirable chemical action in a product; added in small quantity to gasolines to prevent oxidation and gum formation, to lubricating oils to stop color change, and to corrosive environments to decrease corrosive action.

interface *n:* the contact surface between two boundaries of liquids (e.g., the surface between water and the oil floating on it).

ion *n:* an atom that has acquired a net electric charge through the loss or gain of one or more electrons.

***IR* drop** *n:* voltage drop, as determined by the formula $E = IR$.

iron count *n:* a measure of iron compounds in the product stream, determined by chemical analysis, that reflects the occurrence and the extent of corrosion.

K

Kirchoff's second law *n:* the law stating that, at each instant of time, the increases in voltage around a closed loop in a network is equal to the algebraic sum of the voltage drops.

L

lanolin *n:* wool grease, derived from the preparation of raw wool for spinning; it is used in cosmetics, shampoos, and a variety of industrial products.

linear polarization *n:* a technique used to measure instantaneous corrosion rates by changing the elctrical potential of a structure corroding in a conductive fluid and measuring the current required for that change.

lipophilic *adj:* having an affinity for lipids, a class of compounds that includes most hydrocarbons.

M

metal *n:* opaque crystalline material, usually of high strength, that has good thermal and electrical conductivity, ductility, and reflectivity.

metallic circuit *n:* the path of electric current through the metallic portions of a corrosion cell.

mill scale *n:* thin, dense oxide scale that forms on the surface of newly manufactured steel as the steel cools.

Mill scale can become cathodic to its own steel base, forming galvanic corrosion cells.

molecule *n:* the smallest part of a compound that can exist on its own. The atoms of which it consists may be different (such as the hydrogen and oxygen atoms of water, H_2O) or the same (such as the two hydrogen atoms of free hydrogen, H_2). See *atom* and *compound.*

N

NACE *abbr:* National Association of Corrosion Engineers.

National Association of Corrosion Engineers *n:* organization whose function is to establish standards and recommended practices for the field of corrosion control. It is based in Houston, Texas.

nipple *n:* a tubular pipe fitting threaded on both ends and less than 12 inches long.

noble metal *n:* any of the metals with low reactive tendencies at the upper end of the electrochemical series.

nonferrous alloy *n:* alloy containing less than 50 percent iron.

O

Ohm's law *n:* a law that concerns the behavior of electrical flow through a conductor. Ohm's law is stated as

$$R = E/I$$

where

R = resistance, E = volts, and I = current.

The law is used in measuring the resistivity of a substance to the flow of electric current.

ore *n:* a naturally occurring mineral containing metal that can be extracted.

oxidation *n:* a chemical reaction in which a compound loses electrons and gains a more positive charge.

oxide *n:* a chemical compound in which oxygen is joined with a metal or a nonmetal.

oxygen-concentration cell *n:* a corrosion cell formed by differing concentrations of oxygen in an electrolyte.

P

packer fluid *n:* a liquid, usually mud but sometimes salt water or oil, used in a well when a packer is between the tubing and casing. Packer fluid must be heavy enough to shut off the pressure of the formation being produced, must not stiffen or settle out of suspension over long periods of time, and must be noncorrosive.

partial pressure *n:* the pressure exerted by one specific component of a gaseous mixture.

passivation *n:* the process of rendering a metal surface chemically inactive, either by electrochemical polarization or by contact with passivating agents.

permeability *n:* 1. a measure of the ease with which fluids can flow through a porous rock. 2. the fluid conductivity of a porous medium. 3. the ability of a fluid to flow within the interconnected pore network of a porous medium.

persistence *n:* the durability or longevity of inhibitors used in corrosion control.

pH value *n:* a unit of measure of the acid or alkaline condition of a substance. A neutral solution (like pure water) has a pH of 7; acid solutions are less than 7; basic, or alkaline, solutions are more than 7. The pH scale is a logarithmic scale; a substance with a pH of 4 is more than twice as acid as a substance with a pH of 5. Similarly, a substance with a pH of 9 is much more than twice as alkaline as a substance with a pH of 8.

phenolics *n:* thermosetting plastic materials formed by the condensation of phenols (containing C_6H_5OH) with aldehydes (containing CHO) and used for protective coatings for oil field structures.

piston rod *n:* 1. a metal shaft that joins the piston to the crankshaft in an engine. 2. a metal shaft in a mud pump, one end of which is connected to the piston and the other to the pony rod.

polished rod *n:* the topmost portion of a string of sucker rods, used for lifting fluid by the rod-pumping method. It has a uniform diameter and is smoothly polished to effectively seal pressure in the stuffing box attached to the top of the well.

polyester *n:* a thermosetting or thermoplastic material formed by esterification of polybasic organic acids with polyhydric acids.

polymerization *n:* the bonding of two or more simple molecules to form larger molecular units.

positive-displacement meter *n:* a mechanical, fluid-measuring device that measures by filling and emptying chambers of a specific volume, also known as a volume meter or volumeter. The displacement of a fixed volume of fluid may be accomplished by reciprocating or oscillating pistons, by rotating vanes or buckets, by nutating disks, or by using tanks or other vessels that automatically fill and empty.

pozzolan *n:* a natural or artificial siliceous material commonly added to portland cement mixtures to impart certain desirable properties. Added to oilwell cements, pozzolans reduce slurry weight and viscosity, increase resistance to sulfate attack, and influence factors such as pumping time, ultimate strength, and watertightness.

probe *n:* any small device that, when brought into contact with or inserted into a system, can make measurements on that system. In corrosion, probes can measure electrical potential or the corrosivity of various substances to determine a system's corrosive tendencies.

pulse-echo technique *n:* corrosion-detecting processes which, by recording the action of ultrasonic waves artificially introduced into production structures, can determine metal thicknesses and detect flaws.

R

radiographic examination *n:* photographic record of corrosion damage obtained by transmitting X rays or radioactive isotopes into production structures.

rathole *n:* 1. a hole in the rig floor, 30 to 35 feet deep, lined with casing that projects above the floor, into which the kelly and swivel are placed when hoisting operations are in progress. 2. a hole of a diameter smaller than the main hole that is drilled in the bottom of the main hole. *v:* to reduce the size of the wellbore and drill ahead.

rectifier *n:* a device used to convert alternating current into direct current.

reduction *n:* a chemical reaction in which a compound gains electrons and obtains a more negative charge.

residuals *n pl:* the heavy refined hydrocarbons that are used as fuels. Bunker C oil is an example of a residual.

resistance *n:* opposition to the flow of direct current caused by a particular material or device. Resistance is equal to the voltage drop across the circuit divided by the current through the circuit.

resistivity *n:* the electrical resistance offered to the passage of current; the opposite of conductivity.

rod score *n:* scratch on the surface of sucker rod or piston rod.

S

sacrificial anode *n:* in cathodic protection, anodes made from metals whose galvanic potentials render them anodic to steel in an electrolyte. They are used up, or sacrificed.

salt *n:* a compound that is formed (along with water) by the reaction of an acid with a base. A common salt (table salt) is sodium chloride, NaCl, derived by combining hydrochloric acid, HCl, with sodium hydroxide, NaOH. The result is sodium chloride and water, H_2O. This process is written chemically as

$$HCl + NaOH \rightarrow NaCl + H_2O.$$

Another salt, for example, is calcium sulfate, $CaSO_4$, obtained when sulfuric acid, H_2SO_4, is combined with calcium hydroxide, $Ca(OH)_2$.

salt water *n:* a water that contains a large quantity of salt; brine.

sand *n:* 1. an abrasive material composed of small quartz grains formed from the disintegration of preexisting rocks. Sand consists of particles less than 2 millimetres and greater than $^1/_{16}$ of a millimetre in diameter. 2. sandstone.

saturation *n:* a state of being filled or permeated to capacity. Sometimes used to mean the degree or percentage of saturation (as, the saturation of the pore space in a formation or the saturation of gas in a liquid, both in reality meaning the extent of saturation).

scavenge *v:* to remove exhaust gases from a cylinder by means of compressed air. Such removal takes place in all two-cycle diesel engines.

separator *n:* a cylindrical or spherical vessel used to isolate the components in mixed streams of fluids.

slow-release inhibitor *n:* corrosion-preventive substance that is released into production fluids at a slow rate.

soil stress *n:* the uneven penetration of pipeline coatings because of changes in soil volume and moisture along the pipeline bed.

solution *n:* a single, homogeneous liquid, solid, or gas phase that is a mixture in which the components (liquid, gas, solid, or combinations thereof) are uniformly distributed throughout the mixture.

specific gravity *n:* the ratio of the weight of a given volume of a substance at a given temperature to the weight of an equal volume of a standard substance at the same temperature. For example, if 1 cubic inch of water at 39°F weighs 1 unit and 1 cubic inch of another solid or liquid at 39°F weighs 0.95 unit, then the specific gravity of the substance is 0.95. In determining the specific gravity of gases, the comparison is made with the standard of air or hydrogen.

splash zone *n:* the area of an offshore structure that is regularly wetted by seawater but is not continuously submerged. Metal in the splash zone must be well protected from the corrosive action of seawater and air.

spontaneous combustion *n:* the ignition of combustible materials without open flame.

strata *n pl:* distinct, usually parallel beds of rock. An individual bed is a stratum.

stray current *n:* a portion of an electric current that flows over a path other than the intended path, causing corrosion of structures immersed in the same electrolyte.

sucker rod *n:* a special steel rod; several rods screwed together make up the mechanical link from the beam pumping unit on the surface to the sucker rod pump at the bottom of a well. Sucker rods are threaded on each end and manufactured to dimension standards and metal specifications set by the petroleum industry. Lengths are from 25 to 30 feet; diameter varies from ½ to 1⅛ inches. There is also a continuous sucker rod.

sulfate-reducing bacteria *n:* bacteria that digest sulfate present in water, causing the release of hydrogen sulfide, which combines with iron to form iron sulfide, a troublesome scale.

sweet crude *n:* also called sweet crude oil. See *sweet crude oil.*

sweet crude oil *n:* oil containing little or no sulfur and especially little or no hydrogen sulfide.

synergistic effect *n:* the added effect produced by two processes working in combination, greater than the sum of the individual effects of each process.

T

tape wrapping *n:* rolls of plastic sheeting with a preapplied adhesive used to coat buried pipelines in order to prevent corrosion.

thermoplastics *n:* any of a variety of materials often used in pipe coatings, whose molecular structure allows them to repeatedly soften when heated and harden when cooled.

thermosetting plastics *n pl:* plastics that solidify when first heated under pressure, but whose original characteristics are destroyed when remelted or remolded.

time-release *n:* feature built into oil field inhibitors that allows them to be introduced into production systems and their active ingredients released at certain timed intervals.

treater *n:* a vessel in which oil is treated for the removal of BS&W by the addition of chemicals, heat, electricity, or all three.

tube bundle *n:* the inner piping of a condenser or heat exchanger, typically consisting of a group of pipes placed inside the shell of a tank.

tubing *n:* small-diameter pipe that is run into a well to serve as a conduit for the passage of oil and gas to the surface.

tubing head *n:* a flanged fitting that supports the tubing string, seals off pressure between the casing and the outside of the tubing, and provides a connection that supports the Christmas tree.

U

ultrasonic devices *n pl:* corrosion-monitoring devices that transmit ultrasonic waves through production structures in order to locate discontinuities in metal structure, which indicate corrosion damage.

use test *n:* periodic equipment inspection to determine corrosion damage in fields that have been producing for several years.

V

venturi effect *n:* the drop in pressure resulting from the increased velocity of a fluid as it flows through a constricted section of a pipeline.

voltage *n:* potential difference or electromotive force, measured in volts.

voltmeter *n:* an instrument used to measure, in volts, the difference of potential in an electrical circuit.

Self-Test Answer Key

1: The Corrosion Process

1. reduction
2. oxidation
3. base
4. noble
5. corrosion cell
6. salts; acids; hydroxides
7. anode; cathode
8. cathodically controlled
9. T
10. F
11. F
12. F
13. E
14. D
15. C
16. A
17. B

2: Common Corroding Agents

1. Brine
2. splash zones
3. carbon dioxide
4. iron sulfide
5. oxygen
6. oxygen-concentration
7. soil moisture
8. Anaerobic; Aerobic
9. hydrogen sulfide
10. synergistic
11. T
12. F
13. T
14. T
15. F

3: Measurement and Detection

1. iron count
2. linear polarization
3. hydrogen patch probe
4. Hydrogen embrittlement
5. half-cells
6. spontaneous combustion
7. Radiographic surveys
8. pulse-echo
9. (1) inside; (2) outside; (3) isolated; (4) circumferential
10. Calipers
11. T
12. F
13. F
14. F
15. F
16. T
17. F
18. T
19. T
20. F

4: Methods of Corrosion Control

1. system design
2. brine damage
3. alloy
4. interrupt the current flow in corrosion cells
5. changes in temperature
6. (1) soil stress; (2) mollusks; (3) weather; (4) poor handling
7. Inhibitors
8. Lipophilic
9. soluble; dispersible
10. persistence
11. (1) compatibility with other chemicals; (2) emulsion-forming properties; (3) effect of temperature
12. the concentration of inhibitor can be maintained at a constant level
13. squeeze
14. the formation was plugged
15. pitting
16. F
17. T
18. F
19. T
20. F
21. F
22. T
23. T
24. F
25. F

5: Cathodic Protection

1. cathodically protected
2. (1) provide the necessary areas of contact with electrolytes; (2) last long enough to be economical; (3) minimize power requirements
3. Sacrificial
4. end effect
5. minimize the resistance to current flow in the vicinity of anodes.
6. (1) it is used up at a low rate; (2) it provides a uniform environment for the anode; (3) it increases the effective anode size; (4) it reduces costs because it is a low-cost material.
7. Platinum
8. moisture penetration
9. Deep well

10. uneven wear from unequal current discharge
11. Aluminum
12. cathodic surfaces become polarized and resist entry of more current.
13. the electrolyte has a higher electrical resistance than metal
14. the surface to be protected; the surrounding electrolyte
15. $I = E/R$
16. F
17. F
18. F
19. F
20. F
21. T
22. T
23. F
24. F
25. T